Super Clues to Reality

Marian Matthews

© Marian Matthews 2020

Marian Matthews has asserted her rights under the Copyright, Design and Patents Act 1988 to be identified as the author of this work.

All rights reserved. No part of this book, in part or in whole, may be reproduced, stored in a retrieval system, or transmitted in any form or by any means, electronic, mechanical, photocopying, recording or otherwise, with the exception of fair use for the purpose of review, without the prior written permission of the author.

By the same author:
Aspects of Reality – A User's Guide to the Universe
available from
www.marianmatthews.com

A CIP catalogue record for this book is available from the British Library.

ISBN 978-1-5272-5075-8

Edited and typeset by Roma Harding

Cover design by James Wallis
www.instagram.com/jamesmwallis

Printed and bound by IngramSpark

To my husband Toby
and my family.

"I love Marian's enquiring mind and desire to dig deeper into life's mysteries. Her book is like a beginner's reference guide to the universe. She does not profess to be an expert on the subjects she shares, but she is an expert explorer and sharer of insights."

Elaine Harrison, *Author, Coach & PR*

"This book develops subjects from *Aspects of Reality*, taking topics to a higher level, giving powerful insights into worlds beyond our usual understanding. Each chapter makes totally absorbing and fascinating reading. Over many years, Marian has been attempting to make sense of everything going on around her, and to get nearer to the truth about the reality of this world of ours. Her explorations include the works of great thinkers and contemporary science, which she then seeks to reconcile with her own experiences.

What makes her writing so very special is that she takes her readers on a journey with her. Marian selectively dips into the vast store of information she has accumulated to give her readers an opportunity to follow the clues for themselves and reach their own conclusions. She gently guides without ever lecturing. At a time when there is so much uncertainty in the world, she gives us a much-needed opportunity to investigate our very purpose and find some meaning. In *Super Clues to Reality*, Marian continues the magic and delves into such themes as the remarkable achievements of the ancients, reality manifestation, our own consciousness, and the hidden world of other beings all around us, giving tantalising clues to the ultimate big picture of reality."

Richard Raymond, *Writer, Healer, Counsellor*
www.richardraymondreiki.com

Acknowledgements

Grateful appreciation to all the friends I have spoken to and gleaned information from; there have been many of you. For kindly granting permission to include selected extracts from exchanges of personal correspondence, which has been invaluable in my researches, I would like to thank Chris Harris, Sarah Hayward, Matthew Gibson, Judith Lock, Tracey Morfitt and "Primal".

I would also like to acknowledge and thank those professionals in their own field who have provided reviews: Peter Knight, Josephine Sellers, Richard Raymond and Elaine Harrison.

A special thank you to Roma Harding for her clever editing skills and much appreciated contributions, which helped to make this book a coherent whole. Also, to my grandson James Wallis for his vibrant cover design.

And especially to my husband Toby, who has given me the support and space to complete my mission.

What lies behind us and what lies before us
are tiny matters compared to what lies within us.
Buddha

Reality is merely an illusion,
albeit a very persistent one.
Einstein

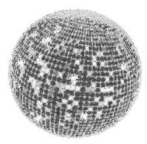

CONTENTS

Chapter 1	Preface		11
	Super Clues embedded in		
	Our Physical Construction		15
	How Did We Get Here?		15
	What Are We and Our Universe Made From?		16
	The Building Blocks of the Universe		17
	What Science Says About Our Reality		18
	Does Human Consciousness Affect Matter?		22
	Alternative Creation Theories		23
	Summary		24
Chapter 2	**Super Clues embedded in the**		
	Nature of Human Consciousness		26
	The Riddle of Human Consciousness		27
	What is Consciousness?		28
	One System or Two?		29
	The Origins of Human Consciousness		30
	Conscious Creation		32
	The Art of Manifestation		33
	Summary		35
Chapter 3	**Super Clues embedded in**		
	Big Science and Cosmology		37
	Cosmology		38
	Alternative Cosmological Theories		39
	The Nature of Time		42
	Summary		45

Chapter 4	**Super Clues embedded in Our Spirit Bodies**	47
	The Spirit Body	47
	The Aura	49
	The Layers of the Spirit Body	50
	Chakras and Healing	51
	The Soul	53
	Source	55
	Summary	55
Chapter 5	**Super Clues embedded in Life After Death**	57
	The Soul Beyond Death	58
	Out of Body Experiences	60
	Ghosts	61
	Disembodied Spirits	62
	Karma	63
	Reincarnation	65
	The Akashic Records	68
	Good and Evil	70
	Summary	74
Chapter 6	**Super Clues embedded in Universal and Earth Energies**	76
	Solar and Lunar Energies	76
	The Earth's Magnetic Field	79
	Gravitational Forces	80
	Leys	82
	Dowsing	84
	Energy Vortexes	85
	Planetary Grids	87
	Gaia	90
	Summary	92

Chapter 7	**Super Clues embedded in**	
	Alien Beings and Other Dimensions	93
	Alien Communication	93
	UFO Phenomena	95
	Types of Aliens	98
	UFO Sightings	100
	Testimonials	103
	Intergalactic Intervention	105
	Ancient Aliens	107
	Stories from Around the World	109
	Lost Technology	114
	Men Ahead of Their Time	116
	Summary	118
Chapter 8	**Super Clues embedded in the**	
	Spiritual & Angelic Realms and Orbs	119
	Angels	119
	The Religious View of Angels	121
	Other Types of Angels	122
	Where Do Angels Come From?	125
	Spirit Guides	126
	Ascended Masters	128
	The Great White Brotherhood	128
	Orbs	129
	Summary	133
Chapter 9	**Super Clues embedded in**	
	Nature Spirits and Elemental Beings	135
	Nature Spirits	136
	Other Fairy Folk	139
	Elemental Beings	142
	Summary	145
	Conclusion	147
	About the Author	153
	Bibliography	155

"All matter originates and exists only by virtue of a force which brings the particle of an atom to vibration and holds this most minute solar system of the atom together. We must assume behind this force the existence of a conscious and intelligent mind. This mind is the matrix of all matter."

Max Planck

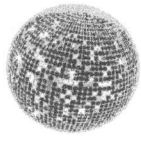

PREFACE

WHAT DO WE MEAN when we talk about our reality? The answer, one might think, is in plain sight. To most people 'reality' means everything we can see around us – our perception of the world and how it works. Science tells us what exists in the greater universe, including the stars and planets in our solar system, even other galaxies beyond our Milky Way.

The truth is, we are unaware of many aspects of our reality that affect us on a day-to-day basis. There is much more we cannot see and accept as fact. For instance, we believe our planet to be a globe, yet we feel as if we are walking around on a flat surface. More surprisingly, the Earth rotates at 1,000 mph and orbits the Sun at 67,000 mph. Yet, we are not aware of any movement at all. What else are we unaware of and perhaps have no idea exists?

In my first book, *Aspects of Reality – A User's Guide to the Universe* (2012), I looked at the science, both large and small scale, as well as philosophical, esoteric and practical, common sense aspects, which all interact to create the reality we think we live in. For my talks, I condensed many of these aspects down to seven major themes and began to find tantalising hints of a much bigger, unseen picture encoded within. Most of us have asked the questions: Who are we? Where do we come from? What are we doing here? Are we the only beings on planet Earth? As it turns out, our world is both more multidimensional and remarkable than many of us imagine.

In *Super Clues to Reality*, I look again at some of the key reality topics in greater depth. Could reality as we know it be something completely different from what we think it is? There are so many clues out there, bound up in what we do know, telling us that what we've been taught does not, and cannot, explain the whole story. There are things going on beyond our visible realms, helping to shape what is around us.

Science has its views, religion has its faith, and alternative thinkers have other ideas, but how do we go about finding out which perspective, if any, is right?

In attempting to explain reality, I use the glitter ball analogy – seen at the beginning of each chapter. As these multi-faceted mirrored globes, hanging from ceilings in ballrooms and discos, slowly revolve, what we see are the reflections bouncing back at us. These reflections look different to each of us, depending on where we are standing and how much light is shining on the glitter ball at the time. So, what we see is our own perception of reality at any given moment. We cannot see into the heart of what we are looking at, only the reflections. It is these reflections that give us clues to the big picture of a reality that is hidden from us.

As science moves on rapidly, many of us acknowledge that it does not explain everything and have asked the same questions I am asking here. Today, humanity is standing on the threshold of a new era, asking questions about the "whys" and "hows" of our existence. It is as if we are poised, waiting for a curtain to rise to reveal hidden depths to the nature of our being, which may change the way we look at the world, forever. If we really are part of a bigger, more complex system than we are aware of, we need to know, and soon, especially given climatic changes and the fragility of our planet, which ultimately responds to how we have been treating it.

As relatively limited human beings and with the tools available to us, we can probably never know the whole story and perhaps we are not meant to. My research identifies what I refer to as Super Clues, which hint at a much bigger picture, beyond current scientific understanding than conventional thinking allows. After all, some of us have skills which cannot be explained by accepted thinking, such as psychic ability and telepathy, for example. Science itself, especially on a cosmological scale, is finding things that just do not add up – more and more cracks in mainstream assumptions do not 'fit' preconceived notions.

Surprisingly, some truths prove to have always been there – in ancient wisdom and Eastern mysticism. We find many of the mysteries of human consciousness to be based on old knowledge, which is still valid today. Although our ancient thinkers did not have the words or understanding of the mechanisms involved, the general principles remain the same. Similarly, we find creation myths all over the world are uncannily similar.

PREFACE

The Super Clues identified in this book are just that – clues. They highlight the fundamental flaws in what we have been conditioned to believe and give indications of other dimensions in our world. We explore what sort of beings we really are, who made us and how we ended up here. In seeking to find out more about our ultimate reality, we look at alternative possibilities and gaps in current understanding. This poses more questions than answers, more insights and conjecture than hard facts. It is, therefore, neither a textbook nor a definitive account, simply what I have identified, as a generalist rather than a specialist, through my own research and study. It attempts to offer signposts to a fascinating, expansive picture of amazing realities that we are not usually taught exist.

These chapters cover topics such as science and cosmology, the soul and human consciousness, the nature of time, our energy bodies, karma and reincarnation, as well as other beings that inhabit our reality with us, including angels, orbs, nature spirits and the elementals. Follow me on this journey of exploration, pushing boundaries and beliefs, to examine the Super Clues embedded in the ultimate truths behind existence itself.

Marian Matthews

July, 2020

SUPER CLUES TO REALITY

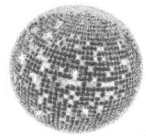

CHAPTER 1

Super Clues embedded in Our Physical Construction

HOW DO WE BEGIN untangling the mysteries of reality? When I began my research, all those years go, I realised that we, ourselves, were one of the best clues we had as a starting point in attempting to understand the true nature of our reality. I asked myself the following questions: Where did we and our planet come from? What sort of beings are we and what are we made of? What are we doing here on planet Earth?

I began to explore the various creation theories and the science behind our physical reality. I then looked at the structure of matter itself and the strange laws governing its behaviour. This provided the first real insights into how things might actually work.

How Did We Get Here?

How life evolved on this planet is, as always, a fundamental question – one which has been asked since the beginning of time. By the mid-19th century there were two accepted but opposing theories. Firstly, was the premise of a creator God and the beliefs that arise from a religious faith. Secondly, the science-based school of thought telling us the creation of the universe and life on Earth was the result of natural evolution. In more recent decades, there have been some interesting alternative theories.

Growing up, along with many others, I was taught that the world was created by God and then populated by animals and plants. We, as human beings, were designed in God's image and put on this Earth for His purposes. There are many cultural variations on how this came about, but mostly, if not always, the principles are similar. There are, however, exceptions. For instance, Buddhism has no creator God to explain the origins of the universe. Instead, it teaches that everything is interconnected; present circumstances are the consequence of past happenings, and present actions become the cause of future events.

I learnt that science believes the universe was created as a result of an accumulation of random natural processes, arising from the so-called "big bang". This is described as an intense burst of energy emanating in an instant from a single point in empty space. The theory goes that dust clumped together to become stars and planets, which formed over a long period of time from these energies as they cooled. Life on Earth began almost by accident, and human beings came about as the result of a process we call evolution. Science, however, does not answer the question of where the energy that caused the "big bang" came from. It is also worth mentioning that not all scientists believe this is what really happened.

Thus, we were either made by God to a deliberate design or came about as a result of evolutionary processes and natural selection, as proposed by naturalist Charles Darwin in his seminal work *On the Origin of Species* (1859). Therefore, even on this basic question of our origins, there is some doubt. Which version should we believe? And does either science or religion tell us the full story? Both, of course, could be true, but equally neither.

What Are We and Our Universe Made From?
What sort of beings are we? Well, on one level we are solid physical entities, living on a rocky planet in a universe warmed by a big ball of gas, called the Sun. This sun is one of trillions of suns in our galaxy. We seem to be revolving in a limitless sea of galaxies and suns. These stretch as far as our most powerful instruments can detect, and beyond.

Human beings on this planet are warm-blooded mammals, bipeds, top of the tree of life, so they tell us. When we look around us, we see a great many animals, plants and fellow humans. We seem to be solid physical entities, existing by the nature of complex biological systems which allow our bodies to function and thrive. We have large brains enabling us to think and communicate, beyond what is needed for mere survival; yet apparently, we use only a very small percentage of its capacity. Is this, however, all we are?

The first set of scientific Super Clues become apparent when we look below the surface and ask what we, and everything else around us, are made of? The simple answer is that every part of the universe, including humanity, is constructed basically from atoms. Every substance in our reality, from the air that we breathe, our blood, bones and biological systems, to the bedrock of our planet, are made up of different elemental atoms in combination. We are our atoms.

The Building Blocks of the Universe
These atoms are the building blocks of everything around us. Not all that long ago, the atom was thought to be the smallest indivisible component in the construction of our physical universe. By the early 1930s, scientists had managed to split the atom. They found a cluster of particles held together by the forces inside it. Each atom was believed to consist of a positively charged nucleus of protons and neutrons, with negatively charged electrons (equal in number to the protons) circling around them.

Every basic element, such as iron or oxygen, for example, is made up of a different combination of sub-atomic particles. Hydrogen is the simplest element, with one proton and one neutron in the centre and an electron circling around. More complex elements have a greater number of ingredients, so to speak, and are held together by the atomic forces within them. The number of protons determine the element's atomic number. This gave us the periodic table, in which the elements are arranged by atomic number, electron configuration, and recurring chemical properties.

When I was at school, any illustration of atoms was shown like a miniature solar system, with a cluster of balls representing protons and neutrons encircling the nucleus, and other electrons whizzing around them. This, of course, was a misrepresentation. The truth is that all atoms, plus other sub-atomic particles that abound in the universe, are themselves made of nothing more than parcels of vibrating energy. These are called quanta (the plural of quantum), which are discrete pockets of vibrating energy. If it were possible to observe an atom, you would not actually see little balls but tangles of vibrating energy, always in motion. These minute vibrating particles take up a very small fraction of an atom. Atoms are mostly just empty space.

Thus we, and all of creation, are constructed from atoms; we are essentially made up of nothing more than parcels of vibrating quanta, held together by the structural forces within them. So, humans and everything else around us are made of energy; we are, in fact, mostly made up of empty space. Yet we seem solid. How can this be? Well, apparent solidity is just a matter of relative density. The chair I am sitting on is physically denser than my body, so I do not fall through it. The air is less dense, so I can walk through it. This gives us our first Super Clue.

The apparent solidity and fixed nature of our universe is an illusion.

Gravitational, nuclear and electromagnetic energies, and those within the atomic structure of matter itself, hold everything together in our physical reality (see Chapters 3 and 6). As energy beings, we and everything around us exist in a mesh of energies. These energies weave together to help form the known universe. We are, then, on one level, no more than energy beings living in an energy matrix, in a quantum sea of light. Even our thoughts and emotions, affecting each of us as individuals, are nothing more than energy. This is a scientific fact. We could, after all, be holograms, or even be living in a programmed universe run by a master computer somewhere! Nonetheless, the truth about our physical construction gives us our second Super Clue.

We and all around us are energy beings living in an energy matrix that is part of the quantum field of light.

What Science Says About Our Reality
Scientists have, over the last few hundred years, formulated laws which govern our physical reality. There is the conventional Newtonian view and the more recent, cutting-edge quantum mechanics.

- **The Newtonian view**
  Our universe is multi-layered. We are told, on the surface at least, that our visible physical world is governed by gravity. Newton's law of universal gravitation defines the force of attraction between all objects that possess mass. The story goes that his big breakthrough happened when he noticed an apple falling straight down from a tree. Newton also came up with the laws of motion, documented in his *Philosophiae Naturalis Principia Mathematica* (1687), which says that for every action there is an equal and opposite reaction. The premise, until recently, has been that all interactions between substances in our complex physical universe happen independently of the observer. The view at the time of Sir Isaac Newton (1643-1727)

was of nature as an intricate, impersonal, inert machine that runs by itself whether anyone is noticing it or not. Every particle within the universe effectively follows the same rules; it always behaves in the same predictable way in the same circumstances. If something appears to behave in an unpredictable way, this is because we do not understand all the rules correctly. The sub-atomic world, however, behaves in accordance with its own laws, which operate at another level and underpin all matter in our universe. This casts doubt on some of the certainties that conventional science thinks it knows.

- **Quantum mechanics**
  Quantum mechanics is the science of the infinitesimally small. It developed over many decades and looks at the behaviour of the sub-atomic particles that make up our physical reality. The word *quantum* comes from the Latin for "how much". QM denotes the minimum amount of a physical entity involved in any interaction. The objects described are neither particles nor waves but a category that share some of the properties of each. Light, for example, looks at frequency and wavelength. It is a very complex, cutting-edge, branch of physics which, in spite of its strange rules, gives us an insight into the way that particles, our fundamental building blocks, can behave. The apparent order of our reality, the Newtonian mechanical rules, are called into question. For instance, if Newton's apple falling from a tree was subject only to the laws of QM, then it would sometimes fall down, sometimes travel upwards or in any number of directions – or all possibilities at the same time. This is known as the uncertainty principle named after Heisenberg, the man who reasoned that the more precisely an electron's position is known, the less precisely its momentum can be predicted from initial conditions, and vice versa.

  There are several quantum mechanics principles, four of which are especially interesting in our search for Super Clues:

  - Sometimes particles are just, well, particles. Sometimes, seemingly randomly, they can become waves. Waves can spread out over large regions of space and time. They can jump gaps. They can sometimes penetrate things thought of as impenetrable. Anything is possible.

- Unlike conventional science which can predict the way that any substance, in a similar condition, will act in the same manner, every time, quantum particles may act in the same way or they may act completely differently. They do not always follow the rules.

- The principle of non-location holds that once particles have been in contact with one another, they are forever linked, even if they end up separated by millions of miles. Given that all energy making up the universe came from the "big bang", every particle has been in contact at its point of origin with every other.

- Individual sub-atomic particles have seemingly endless possibilities of form and position in time and space. Anything that is possible is happening somewhere, all the time. At the moment of transformation between one state and another, they instantaneously take on every possible path. Each particle only settles on one, though, when observed or measured.

To simplify the above, quantum mechanics says:

- The simple act of observing [changes in] what we see, gives an input into creating our reality.
- You cannot determine where anything is, just the probable likelihood. It can be in more than one place at the same time.
- Once a particle has been in contact with another, it is forever linked.
- Most importantly, anything that is possible is happening somewhere all the time.

This tells us the universe is one huge mesh of interlinked energies, and that's pretty much all there is to it! However, it is also possible that quantum mechanics may conceal an even deeper hidden 'reality', defined in my dictionary as:

"The state of things as they actually exist, as opposed to an idealistic or notional idea of them."

There is another energetic force at work in our universe. No particle ever stays at rest, they are always in motion. The universe is not stable or static, but a mass of sub-atomic particles popping fleetingly in and out of existence. The zero-point field, a ground state of energy, constantly interacts with these sub-atomic particles. This not only happens but is measurable. All the elementary particles around us continually exchange energy. So-called virtual particles come into existence, interact with one another, then annihilate each other and instantly vanish. The laws of science give us another Super Clue.

All of physical creation is fundamentally connected on a quantum level. We are not the separate and solid individuals that we appear to be.

Should one still be sceptical, an article by Anil Anathswarmy, 'Reality's Last Stand', published in *New Scientist* (2018), states:

> "Unlike classical physics, which says the world exists independently of observers and observations, quantum theory strongly implies that reality does not exist, or at least cannot be meaningfully described, until it is observed."

The above quotation may seem strange, yet at the same time it makes logical sense. If each particle has endless possibilities and only settles on one option if observed, does this mean that by observing we can physically change our reality? This is an interesting concept. If it is possible to affect the quantum field, in practice this means the consciousness of the observer brings the particle(s) into set being. Put simply, one of the discoveries of QM is that the input of an observer influences the form an individual particle finally takes. This makes us co-creators of our own reality.

Quantum theory implies that reality is uncertain and does not take final form until observed. The act of observing gives us an input into creating our own reality.

How do scientists know this? One way to prove this theory is by the famous double slit experiment, first demonstrated with light in the early 19th century and which contributed to the discovery of QM. It has since been replicated in different forms many times. Basically, photons, or other fundamental particles, are fired one at a time towards a metal plate with two slits, behind which is a recording medium. The particles are recorded according to which slot they go through. When the experiment was unobserved, the particles, in wave form, were found to be going through both slots at once. When observed, they went back to an expected pattern of individual particles, not waves. The results, presented as different patterns on the recording medium, confirmed that light and matter can display characteristics of both classically defined waves and particles – the duality of quantum mechanics. Importantly, it showed that human input altered their reality.

On one level at least, human consciousness can be shown to affect the form of matter in our reality.

Does Human Consciousness Affect Matter?

What does this mean? Well, on one level it can be said that we are all continually creating and changing our own reality. Scientists say this only applies to the quantum world and does not show itself in our physical reality. Can this be entirely true? This leads on to a further question which may illuminate another Super Clue. Can human consciousness, in terms of our thought processes, really affect the form of matter? The answer, of course, depends not just on QM but on the nature of consciousness – what it is and where it comes from. This is the tricky bit, and on one level our consciousness defines who and what we are. Max Planck, the famous German physicist (1858-1947), whose work gained him the Nobel Prize in 1918, made this statement which gives a whole new perspective on the nature of this relationship:

> "As a man who has devoted his whole life to the most clear headed science, to the study of matter, I can tell you as a result of my research about atoms this much: There is no matter as such. All matter originates and exists only by virtue of a force which brings the particle of an atom to vibration and holds this most minute solar system of the atom together. We must assume behind this force the existence of a conscious and intelligent mind. This mind is the matrix of all matter."[1]

---

[1] Science Quotes by Max Planck. https://todayinsci.com/P/Planck_Max/PlanckMax-Quotations.htm

So, what is he saying here? Planck understood all matter to be vibrating energy. The force that holds this energy together as atomic structure and mass can only be generated by a conscious and intelligent mind. Is there any other source of conscious intelligence, apart from ourselves, that may play a part in the creation of our universe? Could it be the mind of God – could the story of creation, as told in the biblical *Genesis*, be correct, after all? Or could it be ourselves operating from a different dimension? Could it even be an electronic computer mind? These questions, explored further in the next chapter, are fundamentally unanswerable, of course!

There may be a conscious force, an intelligent mind of some sort, behind all of creation.

Alternative Creation Theories
What if the "big bang" theory is wrong, and the universe did not simply 'explode' into existence? The dawning understanding of the nature of the profound laws which govern our energetic universe allow for other creation possibilities. These are a totally different set of alternative theories than many of us may have come across. However, the reality of our construction makes any one of these scientifically based theories, which may seem strange at first, more plausible than we may have previously considered.

Explored in more depth in subsequent chapters are:

- The manifestation of our own reality, using consciousness to manipulate the quantum field (Chapter 2).
- A biocentric universe in which life or consciousness generates reality, rather than vice-versa (Chapter 2).
- A holographic universe (Chapter 3).
- An alien generated computer simulation (Chapters 3 and 7).

Could any of the above shed light on the ultimate truth? In this class of theory our physical reality is not solid or set at all. Further, this is only a small part of the mysterious whole.

What do the Super Clues embedded in the nature of our physical construction tell us about our reality?

Summary
The structure of our physical universe gives us a whole new set of Super Clues into our possible origins. The question of who or what triggered our existence is another matter. All we can say for sure is that our fundamental nature, at source, is no more than vibrating energy. We are basically energy beings living in a pulsating energy matrix. All these interacting forces are connected, at some level, in the wider cosmos. Inherent in the sub-atomic matter that makes up our reality, is uncertainty rather than probability.

The question of where this energy came from is, at present, unanswerable. Even scientists who believe our universe began as an eruption from the singularity of the "big bang", cannot be sure from where the initial burst originated. Our universe – encompassing numerous galaxies – may have even come about in ways we have yet to discover. Planetary life is so complex that, personally, I do not think it could be accidental. So, who or what is responsible? A creator God or an intelligent mind cannot be ruled out. There are also indications that human consciousness may have a part to play in the creation of our own reality.

There is, of course, a developing understanding of the quantum field concerning our fundamental construction and existence as physical beings. At the quantum level, it does seem that human consciousness can manifest or alter vibrational energy in some way. The concept of a fixed reality looks shakier than we might think. We might even be holograms or computer simulations, and not 'real' at all!

To recap, quantum theory implies reality is intrinsically uncertain and does not take final form until observed. By observing, we have an input into our own reality. This theory can then be taken a step further. Some people believe we are generating our own reality – collectively or individually – from the multiple, infinite possibilities inherent in the basic construction of the universe.

Some scientific theories indicate this is not as impossible as it might sound (see Chapter 3). The multiverse theory also implies there may be alternative realties existing simultaneously. The biocentric universe operates along similar lines. Rather than consciousness manipulating the existing quantum field, it holds that life or consciousness generate reality. Thus, consciousness precedes reality (this is explored further in Chapter 2). The fundamental question is, then, could it be a God, an intelligent designer, or our own consciousness at a higher level, co-operating in the creation process?

Another theory states that as we are beings made of no more than energy, living in an energy grid matrix, then it is not out of the question that we, and all of creation, could be holograms. We could even be ourselves operating from a higher dimension. Some people fear or speculate that we are living in a computer simulated reality. Why this might be so is a tricky question to answer – unless, of course, the totality of our reality is nothing more than what could be described as a surreal 'alien' computer game?

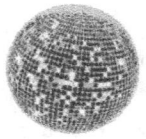

CHAPTER 2

Super Clues embedded in the Nature of Human Consciousness

IN THE FIRST CHAPTER, I explained that we, at the most basic level, are no more than energy beings existing in an energy matrix. We may even be holograms or living in some kind of computer simulation. To make sense of this, we should remember that we are aware of the world around us through our senses. We perceive and glean an understanding of our reality by using our human intelligence. All our interactions with the world around us are channelled through our mind and brain. We only know we exist because we think and are aware of the fact. This is very old knowledge. There is also evidence, at the quantum level at least, that our conscious awareness may affect the form of things around us.

The key to understanding what sort of beings we actually are, is to look at the brain and its capacity for awareness, or consciousness; the ability to directly know and perceive. In this chapter, I will examine this so-called consciousness: What is it? Where does it come from? Why do we have it? This may give us further Super Clues about the true nature of our reality.

The puzzle of the precise nature of human consciousness is one of the mysteries which began my journey of exploring the nature of reality. We take for granted that we have conscious awareness. Yet, what really *is* this uppermost part of our being, which we rely on all the time? Perhaps its purpose is to allow us to know who we are, to be aware of ourselves and the world around us. After all, there is no real evolutionary need for us to have consciousness at the complex level we do, especially if we are little more than mere physical animals.

It will come as no surprise that consciousness is not only very difficult to define but attempting to do so can sometimes lead to contention. Even more interesting is the question of whether human consciousness is a function of the physical brain, a result of the firing of our neurons, or if it is separate from the physical body. So, is it one system or two? If consciousness can be shown to be an extra add-on, so to speak, where could it have come from? And what else does this say about the big picture of our existence – what Super Clues can its very nature yield?

The Riddle of Human Consciousness
It is our physical brain, via our senses, which allows us to know what is going on around us and enables us to interact with our environment. At the same time, it is important to be aware that our senses, as construed by the brain, create an illusion of reality. Information constantly streams into our consciousness to a far greater extent than we are normally aware of. So much input bombards us, we cannot cope with it all. Consequently, our brain sifts all incoming data to look for patterns, which we then interpret as our own version of reality.

What we see through our eyes is a case in point. The thalamus is a large mass of grey matter in the brain which has several functions, including the relaying of sensory signals. Most sensory information is transmitted through here on its way to the outer surface of the brain, the cortex. Brain scans, however, reveal that for all the information coming into the visual cortex, there is six times as much data flowing in the opposite direction. We see and process our perceived reality using only the information that already exists in our heads.

In everyday life, our brains are continually updating patterns based on what we expect to see, rather than necessarily what is actually there. People in conditions of complete sensory deprivation often report still seeing pictures in the form of visual images. Thus, our perception of the world around us relies mainly on what is known already, rather than exclusively on what we are receiving. So, there is a dance between the external world and our brains, the individual internal model we have constructed and everything else. Some of this information we assimilate into our internal schemata, whilst other input gets dismissed or ignored. Ultimately, our vast and powerful subconscious filters what is coming in from our senses to give us our own individual slant or version of reality.

Furthermore, different parts of the brain process information at slightly varying rates. There is said to be a time lag while this happens.

According to David Eagleman, in his book *The Brain: The Story of You* (2015), the reality we see is delayed; it is 2/10ths of a second late. This brain edit we are unaware of. In the end, all we can be certain of is reality is what our brain and to some extent our thoughts, tell us it is.

Our reality is accessed through our brain via our senses. It may not be telling us the whole story.

What is Consciousness?

How do you know you exist, that you are real? Could you be a figment of your own imagination? These may seem strange questions to ask, but the only way we know that we exist as separate individuals is because we think and are conscious of the fact. Even the ancients understood this. As French philosopher, René Descartes (1596-1650) wrote, back in 1637 in his *Discourse on Method*, "I think, therefore I am."

After all, physical sensation only exists through our nervous system, which conducts electrochemical stimuli via the sensory receptors in our brain. However, all could be an illusion. We only know by touch if an object is smooth or prickly, for example, because our nerve impulses send signals to the brain and tells us so. This is similar for sight and hearing, all senses really. Our brain makes sense of what we see and hear, and allows us to experience touch, taste and smell. In terms of sensory perception, awareness is a relative concept. It may also be focused on an internal state, such as a feeling. Pain, pleasure and emotion are also seemingly generated from the bioenergy of our physical brains. This leads to the question, does emotion come from the mind or heart? The only thing we, as individuals, can know for sure is that we exist. We cannot be absolutely certain of anything else.

Put simply, consciousness is the self-aware and thinking part of each of us that makes us human; it is what distinguishes us, we believe, from other animals. Our thoughts are the only 'real' things about us which we all have in common and, as mentioned, this strange and slippery concept is the only way we can be sure we exist. Defining the exact nature of consciousness is the tricky part. Although there is no accepted consensus by various professionals (scientists, doctors, philosophers, psychiatrists, etc.) giving us an exact definition, most cite such things as the ability to have abstract thought, awareness of the past and future, and recognising oneself as an individual entity. These capabilities are, essentially, what makes us who we are.

It used to be thought that animals did not have consciousness in the same way as humans. However, even Darwin recognised that apes and other higher vertebrates, express emotion. These are the reptiles, birds and mammals which lay their eggs on land rather than in water or retain the fertilised egg within the mother. Studies show that higher vertebrates have awareness but do not think in the same way as humans, with an ability to reflect rather than simply acting instinctively. Nowadays, we accept that animals do have consciousness, cold-blooded vertebrates too, even if not to the same degree as us. Even trees and plants demonstrate a level of consciousness which enables them to respond to their environment and communicate with each other, to an extent, for the purpose of survival and to support humanity. It is, however, something notoriously difficult to detect and quantify.

One System or Two?
Two key questions we need to ask is whether consciousness is generated by our physical brains, or is it something else added on? Is it a gift from a creator God, as religious teachings maintain, or has it come about as the result of an advanced evolutionary process, a natural progression of development as we branched off from our ape ancestors?

If it is our mind that makes us human, can we separate mind from consciousness? Our physical brains, our little grey cells, co-ordinate and oversee the running of our bodies, our senses, our muscle control and balance, and all our bodily functions. It controls just about everything, even when we sleep. Our consciousness allows for higher thought, awareness of our place in the world, and a level of understanding beyond the capacity of the physical brain. Is this, then, two separate systems working in harmony, or are they integral?

Where could consciousness have originated from? There are at least two possibilities. Either it comes into the infant's brain at the time of birth, or shortly before or after. Or it is something separate, part of a much bigger picture than most of us are aware of. Maybe a form of consciousness pervades all of creation, in which case what sort of scenario would this entail? The possibilities are mind boggling. If consciousness is an advanced development, the end result of random evolutionary processes which led to us becoming human, that is one thing. If, however, it is something else added on, then it becomes a Super Clue to a much bigger picture. Maybe, however, something completely different is going on.

The human brain is so complex, how it really works is not fully understood. There is ongoing scientific research into the concept of separation of human consciousness from the physical brain. Yet, what evidence is there? Doctors and surgeons who treat people at the point of death often have a keen interest in this phenomenon, which may indicate that we go on, in some form, after physical death. Until now, the main evidence for the brain and conscious mind being two systems rather than one, is substantial but anecdotal. Many people have tales to tell or have had experiences which confirm the conscious part of us can, and does, leave the physical body. These include near death and out of body experiences. The belief that consciousness is part of our soul and returns to something called Source after physical death, before reincarnating into a different body, is explored further in Chapters 4 and 5. If this is the case, it seems, therefore, to be two separate systems.

Human consciousness can be shown to be separate and separable from our physical bodies. Rather than being just one system, mind and body interact and work in unison. Together with what is called our soul, this is what makes us human.

The Origins of Human Consciousness

If consciousness is not just an accidental by-product of the evolution of our physical bodies, where could it have come from? Has it been put or planted there for a purpose? There are several views on this, all of which reveal intriguing possibilities for an alternative picture of our existence. Which is the correct one is, of course, another matter. Current theories, which are not mutually exclusive, are as follows:

- Consciousness is a gift from a God, the universal mind or an intelligent designer of some sort.

- It is incorporated into humanity, at, or around, the time of birth. With consciousness comes Free Will, the ability to make decisions beyond our genetic imperative, meaning souls can be tested. We can be punished or rewarded or simply learn life lessons. This depends on the religious or spiritual view adhered to and assumes the faith-based creation theory to be the correct one.

- There is a universal, collective consciousness all around us. This accepts that consciousness is another fundamental force permeating creation. It assumes that when we are born, we access part of this collective consciousness for the duration of our lifetime. Within us are other layers and depths which can only be accessed by philosophical study and meditation.

- Consciousness permeates every particle of the universe. The accepted cosmological model is of an infinite universe, whereas the observable universe is finite. There is no real scientific explanation why atoms hold together and create the matter of which we and everything else are made of. Thus, there must be an advanced cosmic mind behind creation, whether it is the mind of God or another form of intelligence.

- It is part of Source. As souls, our lives are part of a much bigger universal picture, connected to a point of origin from where we came into being. The principle is that we choose to incarnate in physical bodies to learn and grow. We then return to Source after each lifetime.

- Biocentrism sees inherent value in all living things; life and consciousness are fundamental to the universe, rather than the other way around. This theory maintains some form of intelligence existed before physical creation. Could it be a God or an intelligent designer, even ourselves at a higher level?

- Global Brain Theory asserts that our individual consciousness relies on neurons in a global brain; if we work together, we can collectively change the world around us. Meditation can help us achieve this.

Could all or any of these theories be true? Human consciousness is unquestionable, but its true origins and place in the universe remain a matter of much debate. Nothing can yet be proven. Only time will tell. Apart from the first possibility, that human consciousness is a gift from God, these theories assume a much bigger, more mysterious, picture of reality than we have been taught exists.

Conscious Creation

There is a belief, in some circles, not only that human consciousness can be viewed as a separate entity from our physical bodies, but it is so powerful that it can be used as a creative force. Given we are thought to manifest our own reality, the idea that collectively we may have a role in creating the universe is a difficult yet not unreasonable concept to grasp.

David Russell, who organised a weekly meditation for world peace 'The Global Brain Project' (until his death in 2016), shared with me his view on creating and shaping reality:

> "Reality emerges as a somewhat random series of impulses from a field that consists of nothing but possibilities. This field of pure, unadulterated and unmanifest consciousness is actually the Unified Field of quantum physics. It is a field that exists in all things, but it is liveliest in human beings and animals. It is the shared experience of this field that creates and shapes reality. We all contribute to shaping reality. However, the reality we create takes two forms. Whilst we create a reality that can be shared and is common to all, we also shape a reality that is unique to each of us as individuals. The world as we know it is a tapestry woven from the threads of a reality that we all know, held together with an endless number of individual realities."

So, although we may think we are all living in the same reality, we are in fact generating our own individual realities within the quantum field. We may ourselves be intelligent co-creators, as suggested by Dr Robert Lanza in his book co-authored with Bob Berman, *Biocentrism* (2010), which explores how life and consciousness are the keys to understanding the true nature of the universe. This may seem a far-fetched idea at present. However, in support of this view, Peter Russell presented a vision in *The Global Brain* (2007) of the potential for humanity to become fully conscious super-organisms in an awakening universe. Thus, science may ultimately prove to agree with the following Super Clue.

Human consciousness may be used to affect or to help create or change our reality in an awakening universe.

The Art of Manifestation

Since time began, there have always been those who believed they could alter or enhance their reality. Manifestation is about using the conscious mind to change the reality around you in some way; that it is possible to change things to one's advantage, or to the advantage or disadvantage of others, using prayer, magic or incantation. How this works depends again on one's belief system or view of the nature of reality. Some modern thinkers believe it can be done by the force of sheer intention. Quantum mechanics, with its axiom that everything possible is happening all the time, the observer having input into what actually happens, supports this possibility.

If reality is as fixed as it seems to be, then surely manifestation would not be possible, would it? People throughout the ages, in various ways, believe they have achieved just that. Priests, magicians, witches, and a proportion of humanity adhering to no particular group or allegiance, have always sought intervention to support their own ends or efforts to help others. Manifestation is itself, neutral. Whilst it may be used for good or bad, it is a force beyond judgement.

There are a number of different forms of manifestation:

- **Manifestation using prayer or spells**
  Both the conventionally religious and the magical rely on appealing to intervention from a creator god or goddess, saint or avatar, to achieve their ends. The religious use prayer, which is believed to strengthen one's relationship with God. Wicca invoke spells and incantations. Are they appealing to outside forces, using quantum energies to change reality, or simply focusing mind energy to achieve their goals? Is there a god/goddess out there who would intervene on their behalf? As I understand it, Wicca uses the power and energy of the natural world and the elemental forces to enhance and strengthen any intent or spell. They may also use symbols and ritual. This is particularly powerful in a circle or group.

- **Manifestation by intention**
  Today, the focus on intention has developed even further. There is a school of thought that suggests you can change your own reality; you can manifest the life you want by something as simple as intention. Life on Earth is both complex and interactive; our minds and memories sift out the millions and trillions of bits of information the

world bombards us with. We create, to an extent, the reality we are already aware of. It may be just a matter of putting the focus on what we want and visualising achieving it, for the path to become clear. This is accepted as manifestation on a basic level. Many people believe that we should always ask for the highest good. In doing so, we accept there is a higher power beyond ourselves.

- **Manifestation using the law of attraction**
  There is a belief, in some circles, that it is possible to design the life you want – to generate wealth, relationships and anything wished for – through the power of attraction. Simply put, if you send positive thoughts concerning your desires out into the universe, and imagine you already have what you want, it is possible to attain what you ask for. This concept of the law of attraction has been around for thousands of years, used for various purposes and kept secret by the elite of all societies. Even today, 5% of people control 95% of the world's resources and wealth, and not entirely through economics.

- **Manifestation using quantum physics**
  Quantum physics tells us reality is a projection of our collective consciousness. This being the case, it should be perfectly possible to manipulate reality and create scenarios to achieve the desired outcome. It may be, though, that by sending our intentions out into the universe, we are altering the quantum field and changing reality by our way of thinking. On the other hand, if it is our awareness that creates our reality at any given moment, then changing our reality may be as simple as changing our perception of what is going on around us. In a collective sense, this does, of course, have implications for so-called brainwashing, as well as working together for the good of humanity.

- **Manifestation using the multiverse**
  In Chapter 3, I discuss the multiverse. This current scientific theory states that for every decision we make, we split, so there are multiple copies of each of us in multiple realities, carrying out every possible variation of our actions and making decisions based on all possible options. If we are unhappy with our reality, we can, in theory, just by asking, change to another line or go through another doorway.

This not only touches on the theory of parallel universes, which branched off from ours (or vice versa) and enables life to be played out in alternate realities, it is, at a higher level, a revolutionary pathway to manifest change and connect with Source (Chapter 4). This also relates to the Buddhist concept of mindfulness, a spiritual practice recognised by the Ascended Masters (Chapter 8).

Some people believe they can alter reality and change expected outcomes by manifestation, using the mind or appealing to a higher power. If tapping into the quantum field is possible, then reality must be more fluid and multidimensional than it appears to be.

What does the concept of manifestation mean for our reality? If changing reality by thought, prayer, spells, or intention is possible, then what is the mechanism? How it is done would explain a lot about what constitutes our reality. If we are using the quantum field, we must be dealing with the uncertainty of matter. If, however, we are relying on a higher power or the possibility of parallel dimensions, then we are looking at a deliberate, purposeful and structured universe.

What do the Super Clues embedded in the nature of human consciousness and the art of manifestation, mean for our reality?

Summary
At present, our understanding of where soul consciousness comes from and may ultimately go after death, depends on a combination of personal belief and experience, together with a level of understanding, which some would say requires a depth of spiritual practice. The big questions posed by its very existence are the presently unanswerable ones. Why, as mere animals, are we conscious at all? Why do people choose, assuming they have any say in the matter, to incarnate on planet Earth? The nature of human consciousness gives us a clue, but, until we know more, the answers again remain elusive.

If we can alter or expand our personal and global reality using our consciousness, or by manifesting what we choose, this would open-up multiple possibilities. For instance, we could be gods, creating reality to test ourselves, playing an interactive game in some kind of cosmic playground. Then again, perhaps we are we not meant to know such things and, as human beings, are simply here to enjoy a physical life of sensory stimulation. At a soul level, however, it is important to remember we are fragments of our universal selves, with the ability to reconnect with our Divine purpose.

All I can say is, the nature of human consciousness is perhaps the most important Super Clue to our existence out there. We know that consciousness exists and now, more than ever, we need to embrace its potential towards creating positive collective and individual realities for the future. It is only through consciousness that each of us know we are alive, as 'real' human beings living in Earth reality.

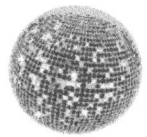

CHAPTER 3

Super Clues embedded in Big Science and Cosmology

IN THIS CHAPTER, I LOOK AT the physics of the greater universe to see what we may find if we look closely enough. For all we have already discussed, inherent in the science of the very small – known as quantum mechanics – there are also Super Clues rooted in studies of the big picture of the universe. This is cosmology. The construction and interaction of the stars and galaxies are part of our physical reality. These are the objects visible in the heavens, as well as those which can only be seen with the most powerful of telescopes, optical or radio. It also considers those things invisible to the human eye we know, or suspect, are there.

The accepted scientific theory of our origins used to be that everything began when the random energies of the "big bang" cooled down and became matter. Matter from the surrounding dust clumped together to form stars and then planets, which themselves slowly cooled and, in the case of Earth, developed life in the form of single-cell microorganisms. This cellular life, believed to have first appeared almost four billion years ago, just a few hundred million years after the formation of the Earth itself, then evolved into plants and animals, and eventually human beings, who themselves have evolved over the millennia. Was this how it really happened? Well, some areas of science are no longer entirely sure. As we shall see in Chapter 7, we may even have started out as cosmic hybrids. Interestingly, there are many other credible, but differing, models of reality currently under investigation by leading scientists. New ideas seem to emerge with some regularity. These candidates are worth looking at for Super Clues.

The truth is that our reality is so complex and multi-layered, when we look at the construction of the universe and creation theories, there are more gaping holes and disagreements among academic factions than one might imagine. However, even though not yet entirely understood, sometimes mainstream science can give us real insights into the way things work, without their explanations necessarily needing to tell the complete story.

Cosmology
More and more powerful telescopes, probes and satellites are reaching into the depths of outer space, trying to make sense of both the seen and the unseen. For most of us, things seem to get more complicated rather than easier. There is a limit to how far we can see into deep space, which is a factor determined by the speed of light. All we can see with optical telescopes is light as it travels towards us. As we explore further and further into the depths, we are essentially looking back in time. When we observe the stars in the night sky, what we are seeing, aside from our sun, are all light-years (the distance light travels in one year) away from Earth. There must have been a time, before the "big bang", when nothing was there. Beyond our range is the great unknown – the universe could end there, or it could go on infinitely. At this point in time, we do not actually know. We can only know about the things, perceptible or imperceptible, which our sense organs and instruments are able to detect. There may be other things in our immediate environment that because we cannot see or identify them, we have no idea exist.

The other thing to mention, when exploring what do we really know, is that it is now believed we cannot see 95% of what should be within our range. We can only see what is emitting light or radiation of some sort. Some of this missing 95% is called dark matter. It is not visible to the human eye and, to date, undetectable by modern scientific instruments. Dark matter was first discovered by scientists noticing phenomena such as stars not behaving as expected. So, the thinking is that something else must be there, to explain its behaviour.

Cosmologists now believe the gravitational effect of dark matter is holding the known universe together. Conversely, there is another unseen energy pushing the universe apart. This is known as dark energy. All the galaxies in the universe are moving apart from each other, not just from the central point of the "big bang", so everything is spreading out as if the fabric of space itself is stretching. The point here is there is far more unknown than known about our physical universe. Many theories are being researched and debated and, if any of these are eventually proven to be true, it would take our understanding of the big picture of ultimate reality to a whole new level. Science is trying to make coherent sense of our greater physical universe, thus reality itself.

How can the scientists work things out? Well, physics is a way of putting into mathematical equations all the processes and forces (such as gravity and mass) which are involved in creating and controlling our universe. Physicists are searching for something called the Grand Unified Theory, which would, apparently, tie everything up into a coherent model. There is also the problem of reconciling Einstein's theory of general relativity with what goes on in black holes. Some processes they seem to think are now completely understood; others they are not so sure about. If they cannot make the equations work together as a unified whole, then what are they missing? What would make their calculations hang together? This search for answers has opened-up areas of research into sometimes radical and mind-bending solutions. It has certainly generated some interesting alternative theories, which it is hoped will answer puzzles in relation to the formation and construction of the cosmos.

Alternative Cosmological Theories
There are several interesting theories which, to make the equations work, propose that our universe must be multidimensional.

- **String theory**
  This is a very complex theory which tries to explain certain phenomena that are currently not resolved by quantum physics. Put simply, all the different basic particles that make up our universe vibrate at different frequencies, like notes on a cosmic string. The tricky bit is understanding for the vibrational models to include all the different particles, there must be other dimensions in space for them to vibrate in. The number of required extra dimensions varies, although the theoretical solutions suggest, depending on which mathematician is correct, 13-ish. Some theorists believe the total to be as low as 10, while others feel it could be as many as 26.

- **Missing gravity**
  This looks at the fact that gravity is simply not as strong as theoretically it should be. Why can we manage to lift our feet from

the Earth in such a weak gravitational field? The assumption is that gravity could be leaking into other dimensions. If so, then where does it go? Well, it may be passing over or migrating into existing dimensions which we are unable to see or detect. This further indicates we could be living in a multidimensional universe.

- **Brane theory**
  This proposes that all particles and forces are confined in our universe by a membrane. Only gravity seems to escape. The thinking behind it is there may be other membranes, containing other universes around us, which we are simply not equipped to detect.

There are also other interesting cosmological theories which seem to leave conventional science behind.

- **Multiple universes**
  This is another theory concerning the possible simultaneous existence of many universes. It relates to the existence of not just other dimensions within our universe, but more than one created universe along with our own, which could possibly operate within different laws of physics. They could be totally separate from our reality or we may, unknowingly, switch between them.

- **Balloon universes**
  This theorises having spawned our universe via the "big bang", the "bang" just kept on going – the universe carried on expanding. Other universes were similarly spawned and are somewhere out there, unseen.

- **Multiverse theory**
  This theory simply states there are parallel universes with different realities, existing alongside ours. As we journey through life and make daily decisions, our lives split between the disparate realities for every action we take. So, in deciding between x and y, there will be a reality in which we do x and another reality in which we do y. These will split again on the next decision. Thus, parallel lives are generated in which we live alternate versions of our lives. This is, surprisingly, not a new concept, but one theoretical physicists were discussing over forty years ago. If the multiverse theory is true, it

would mean there are multiple, or even infinite, versions of ourselves, each carrying out every combination of possible actions in multiple realities.

- **The holographic universe**
  The holographic model of the universe is a serious scientific theory being researched worldwide. A hologram is a way of using light to generate a three-dimensional image from a two-dimensional photographic plate. Although one may think we cannot be holograms because we can see and touch what we perceive as a solid world around us, this is not necessarily the case. We may all be living in a 3D movie! As mentioned previously, solidity is an apparent illusion. It is only our sense organs: our eyes, ears, mouth, nose and skin, which lead us to believe in a fixed world. All sensation and feeling are generated in the mind. If we are holograms with consciousness coming in from elsewhere, then we would not necessarily be aware of the fact.

Many cosmological and scientific theories assume a multidimensional or alternative universe. We are, almost certainly, part of a bigger, more complex reality than we can see or easily detect.

Are any of these theories plausible or even likely? Who knows, but they are certainly interesting, if fundamentally unprovable. More and more scenarios are being identified all the time. If any prove to be true, this would be a massive Super Clue to a vastly overarching grand plan, a reality way beyond the grasp of human understanding. Don't forget, limited by the range of our sense organs and technology, there may be much around us we cannot see or detect, which may throw a whole new light on the nature of reality.

We know many species in the animal world have enhanced senses compared to those of humans who rely largely on sight, allowing them to interpret the world in a different way. For example, a dog's sense of smell is about fifteen times more sensitive than our own, whereas a cat's five senses have evolved to allow them to hunt effectively at night. A dolphin's auditory nerve supply is about three times that of humans, possibly providing it with a highly developed ultrasonic system, used for communicating and locating things in its environment.

As ever, this all leads to more thought-provoking questions than answers. The multiverse theory, involving multiple copies of ourselves, alludes to a puzzling and complex reality, beyond most people's grasp. If true, then what happens to our consciousness – can it actually split? Can we jump from possibility to possibility, known as quantum jumping? Some people think we can. This, then, leads to further questions. Are all individual copies conscious of themselves? And what are the implications for the duality of good and evil (discussed in Chapter 5) if copies of each of us encompass every possible variation?

String, Brane and multiple universe theories propose there are many dimensions in our reality, even entire further universes. The holographic universe assumes a whole new level of reality from which the hologram is created, also that there is a place or dimension from where our consciousness originates. If the holographic theory of reality is true, this gives us a particularly juicy set of Super Clues.

Our energetic physical construction, and that of the universe around us, means we could easily be holograms.

Another perplexing thought comes to mind. If it is true that we are holograms, then who and where is the cinematographer and where are the originals? Could they remain with God or an intelligent designer of some sort? As mentioned, could our reality be likened to a complex computer-generated model? If so, where are the programmers? Alternatively, could we be our own co-creators from another dimension?

As if in response to this question, I came across a book written by cosmologist Jude Currivan, *The Cosmic Hologram* (2017), in which she explains how Einstein's theory of relativity and quantum mechanics can be reconciled if we consider energy-matter and space-time as complementary expressions of what she terms 'in-formation'. The author explores how consciousness connects us to the many layers of reality as both co-creators and manifestations of the cosmic hologram.

The Nature of Time
No serious discussion of conventional theories about the structure of the universe can ignore the nature of time. Time is just time, you may say, what is there to discuss? It is not so simple, however. It may or may not surprise you to know there has been an ongoing debate amongst scientists for many years, about the nature of time itself. Does time actually exist? We may think of time as just seconds, minutes and hours

of our lives, passing by in an ordered, fixed stream. We certainly perceive it that way in our daily lives. It turns out, however, that this concept is questionable and may simply be an illusion.

There are two main opinions regarding the nature of time. Sir Isaac Newton believed time was absolute and linear, and marches on regardless of the machinations of the universe. Einstein thought that Newton's view was wrong, and believed time was simultaneous or variable. Einstein saw time-space as part of an interwoven fabric; the presence of matter changes both, stretching the fabric like a weight suspended on a sheet. He wrote,

> "Physicists believe the separation between past, present, and future is only an illusion, although a convincing one."[2]

So, is time linear or does everything happen at the same time, and we just perceive it to be progressing in an orderly, fixed manner? The answers to these hotly debated questions may give us further Super Clues. I was surprised to find what is often considered an alternative view, that everything happens at once and linear time is simply an illusion, to be, in some fields, a mainstream view.

Others feel that time is a fundamental force of the universe. Yet, without understanding how it works, too many of the big questions of physics are left unanswered. On a day-to-day basis, we feel as if we are moving through time in a linear, orderly manner – we perceive time as a straight line. Minutes, hours and days pass with no apparent variation. Indeed, it would be impossible to cope with life in any other way. Things decay and change or come into existence around us; we experience the changing seasons, and, of course, our own ageing process.

In the laws of physics, entropy, a property of thermodynamics, is defined as particles that behave in such a manner as to give the impression of disorder or randomness. Entropy indicates the degree to which a given amount of thermal energy (heat and temperature) is available. The greater the entropy, the less available energy. According to the second law of thermodynamics, while energy cannot disappear because of the law of conservation, the entropy of the universe tends towards maximum. This is a complex theory, not easily understood.

---

[2] Written by Einstein in a letter to Besso's family, when his friend died. http://everythingforever.com/einstein.htm

It is often pointed out that we do not know why these laws are as they are, or why the universe should have started in the way it did. This may seem to be an academic debate, but the principle does matter to us. It has massive implications for our journey through life and our assumptions of human free will. Although this makes little difference to our everyday lives, for the big picture of reality the truth about the nature of time is very important. It may lead to a Super Clue.

If linear time is an illusion and everything is happening all at once, then the future may be fixed, in which case it is predetermined. This means our free will may be called into question. Could this explain why some people can see into the future, because extrasensory perception (ESP) allows them to tap into what, in a sense, has already happened? If, however, time is linear, then we cannot predict what the future will hold. Our decisions at any given moment can, therefore, change the future. Only time, in both senses, will tell us which theory holds true.

Another question that needs to be asked is, how can time be simultaneous and at the same time appear to be linear? To get our heads around this puzzle, the easy answer is that our physical bodies are programmed, or have evolved, to perceive time as linear. The part of our consciousness which makes us what we are as human beings, perceives it to be so. In the space-time continuum, our perception of time as linear, when all may in fact be happening simultaneously, could indicate an element of deliberate design. This is a Super Clue to a designer reality. It is also worth remembering that our sense organs do not necessarily register all that exists around us. Until we understand who we really are and what we are doing here on Earth, we perhaps cannot make sense of the conundrum of the nature of time.

The fundamental nature of time, whether it is linear or just perceived to be so, is a matter for debate. In our seemingly fixed world, this is another aspect we cannot be certain of.

So, what does the nature of time tell us about our reality? We have seen that quantum mechanics describes objects as waves of possibility which can be in two or more places at once. Physics has demonstrated that an atom can simultaneously exist in two separate places. So, in principle, two "entangled" objects can respond instantly to one another. Does this mean that instantaneous communication, quantum leaps of some kind – quantum jumping – must exist?

Shrödinger's Cat is the famous thought experiment, devised in 1935 by Austrian physicist Erwin Shrödinger, to explain what he saw as the flawed interpretation of an object existing simultaneously in all possible configurations. He wanted people to imagine that inside a sealed container was a cat, a miniscule amount of radioactive material, a Geiger counter, poison (hydrocyanic acid), and a hammer. If the Geiger counter detected radiation, the hammer would smash the poison and the cat would die. However, until someone opened the container, it was impossible to predict the cat's outcome. This meant until the system collapsed into one configuration the cat would exist in a paradoxical state of being both dead and alive. This being impossible, Shrödinger claimed that the theory of quantum superposition was flawed and could not work with large objects. Modern experiments have since proven it does work for very small things, like electrons. So, is it only when you observe something, you know it is there?

Chapter 6 touches on the variability of time and how it is affected by gravity. According to Einstein's theory of general relativity, time is part of the interlinked and interactive forces that make up the universe. If time is simultaneous, and past, present and future are happening all at once, this throws up many more questions. Has everything taken place already? Could time be changeable? Is it in an active or static state? And how does Einstein's theory of general relativity affect our perception of time in a practical sense?

So far, we have not mentioned the nature of time in relation to man as a spiritual being, which we shall come on to in the next chapter. As beings inhabiting a physical body, we only project a small fragment of our consciousness into the third dimension. Thus, it is possible that other fragments or aspects of our consciousness exist simultaneously in other dimensions. Is this what we mean by our ascended selves?

What do the Super Clues embedded in science and cosmology mean for our reality?

Summary
In the first chapter, when I explored Super Clues inherent in our physical construction, I discovered we are essentially energy beings – beings of light – living in a vibrating energy matrix.

All the possible scenarios outlined here, including the nature of time and consciousness, offer different sets of Super Clues, either hinting at a designer universe or multidimensional realities, perhaps both. There may even be aspects of ourselves co-existing beyond the third dimension. Which theory, if any, is the correct one, we simply do not know. All these theories hold within them possibilities of different types of reality, currently being researched and argued about by scientists and cosmologists. Even Nobel prize winners seem to disagree.

Perhaps their theories are not necessarily opposing, after all, given the interactive forces to which Einstein refers. Quantum physics has been asking questions about the nature of reality in much the same way as Eastern mysticism, thousands of years ago. All, however, indicate a much more complex, multidimensional picture of the cosmos and space-time than we are taught exists.

The fact there is much uncertainty raises more questions than answers. All we can say for sure is that the concept of time is neither as straightforward nor as predictable as we may believe it to be. Some researchers argue that space and time, matter and energy, are all one; everything is interlinked and our scientific theories are simply an illusion created by our own minds and machines wanting to segregate and categorise everything. Chaos and Complexity Theory might even be man-made. There could be some truth in that!

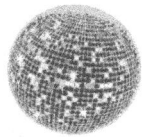

CHAPTER 4

Super Clues embedded in Our Spirit Bodies

THIS TOPIC IS PRESENTLY beyond the scope of science, but is vital to our understanding of who, as members of humanity, we truly are. As we have seen, our biological systems, as seems to be true of all organic entities, are multi-layered. On one level, we have our apparently solid, physical bodies, but at the most basic level we are no more than vibrating parcels of energy. There are layers between these two extremes though, different components to us. One of which is our so-called spirit or subtle body.

So, what does the existence of the spirit body mean in terms of the sort of creatures that human beings really are? Embedded in the concept of spirit are indications of a higher level of self, beyond mere physical survival. People often refer to human beings as having three main aspects: mind (consciousness), body, and spirit. It is the existence of this spirit body which leads us to an important Super Clue in the complexity of our human existence.

The Spirit Body
The human organism has very complex, intricate systems. Our brain, heart, lungs, and other organs, together with our digestive, reproductive, endocrine and nervous systems, must all work in unison as a cohesive whole. If anything goes wrong with the balance of our physiological systems, we have an accident, or are invaded by a virus or bacteria of some kind, then we consider ourselves unwell and the body struggles. In extreme cases of dis-ease, the physical body can indeed shut down and cease to exist as a living entity.

When we look at how complex our bodies are, the wonder is that it ever functions as well as it seems to, for most people. How is everything coordinated? Well, the conventional view is that the physiology of the brain seems to do most of the work. This is not necessarily the whole story though – something else does seem to be going on.

The understanding is that we have, alongside our physical body, a spirit body, sometimes referred to as the subtle energy body. This too is made up of vibrating energy, but of a higher and finer frequency than our physical body. The spirit body supposedly assists and helps co-ordinate our physical systems; the two working together in harmony to maintain our health and well-being. There is a cross-over, however. Any distress or imbalance from our physical body will affect our subtle body, and vice versa. For optimum health, both need to be in balance, separately and with one another.

Some people believe all living things are born with a spiritual blueprint which develops a physical body to its specification, rather than the other way around. As we are so complex, this does make sense. Anyway, let us look at the fundamental question concerning where the spirit body could have come from. There are, of course, differing views as to what it is and where it could have originated from. Could the spirit body be part of a random evolutionary process? I think this is unlikely, as its very existence seems to hint at deliberate design. This is especially true if we all grow to a blueprint. Whilst one could imagine a natural progression of simple organisms developing into more complex beings, and then as humans, the spirit body does not seem to be an entity which could have evolved in this way.

The most illuminating insight was explained to me in 2018, by Sarah Haywood, a friend who works with angels and Ascended Masters:

> "Your spirit in this lifetime is an aspect of your soul. It holds your present consciousness; each time you incarnate that seed is planted."[3]

So, one can imagine a spirit seed being planted, and within it the blueprint of a life form whenever new physical life begins.

As human beings, we have at least three different types of energies within us, all of which interact with the energies of Earth. We have biochemical reactions generated in the running of our physical bodies, we have the energy of our thoughts and feelings, and we have our spirit

---

[3] See also the website of Sarah Haywood, www.aspire2bfree.com/

or subtle bodies. It is the spirit body that helps us to co-ordinate our physical body and link with the Divine. There is also a belief that every living thing has a spirit body. The evidence for its existence arises from direct observation and experience, rather than being scientifically proven. Importantly, all these layers can be seen in our aura.

The Aura

Traces of the subtle or spirit body can be seen in the aura which surrounds every living thing, including trees and plants. The aura is the egg-shaped energy field that can be detected around the physical body, but in fact extends way beyond. Healers and those individuals with a heightened sensitivity can work on the aura when giving healing. Some people can see auras in vibrant colours, others just experience or feel them as vibrating energy fields. The aura holds not just the traces of biochemical activity generated in the running of our bodily systems and brains, but also the overlap of our spirit bodies. Some people believe that a record of our past, present and future – the experiences and life lessons that have happened and will happen to us – is stored here.

Why can we not all see auras? Well, my understanding is that it depends on the quality and sensitivity of one's sensory organs. We all see things differently. The rods and cones in our eyes that allow us to see all vary slightly, depending on our genetics. Just as we all see colour in our own way, the brain interprets what we do see. Further to this, our subconscious edits and filters what we receive and shows us what we expect to see.

Just as some people are gifted with an ability to see things beyond the range of most of us, some people are born with the ability to see auras; sometimes it has developed. For others, it seems to be a gift which those who operate from a spiritual dimension are blessed with.

I myself saw, but only once, a human aura. I was out walking at a beauty spot in Dorset, when my more sensitive friend exclaimed, "Look at that!" I turned around to see a fellow walker with a magnificent aura of green and gold, stretching all around him and above his head, up into the sky. I was not expecting to see anything. It was only a glimpse, but I have never forgotten the experience. I do sometimes notice vibrating energy around some people's heads.

More importantly, proof of the existence of the spirit body may lie in the healing work of many people, using spiritual energy as a tool. A technique of photographing the aura, known as Kirlian photography, officially invented in 1939, shows a multi-coloured silhouette around an object. The image can then be 'read' by a trained person and claims to give an insight into a person's spiritual, emotional and physical health. Imbalances in the aura or spirit body show up long before manifesting physically in a person.

The Layers of the Spirit Body
People across many cultures and belief systems perceive the various layers of the spirit body but may name them differently. The ancient Egyptians have the Ba and Ka – spiritual aspects of an individual believed to survive after the death of the physical body.

What I find the most useful is the following description of the first seven layers:

1. **Etheric body**: aligns with and is the heaviest and closest to the physical body.
2. **Emotional body**: holds emotions and feelings – part of our consciousness.
3. **Mental body**: thoughts, ideas and thinking processes – also part of our consciousness.
4. **Astral body**: the link between those layers bound up in our physical body and at soul level – connected by a spirit cord.
5. **Celestial body**: the spiritual, emotional plane.
6. **Ketheric body**: holds all the other bodies – the mental level of the spiritual plane.
7. **Etheric template**: exists before the physical body – perhaps the super blueprint?

It is the first three layers or levels, the etheric, emotional and mental bodies, that healers usually work on to bring balance to the physical body. The existence of the etheric, celestial and ketheric bodies, and indeed the astral, assume a layer beyond our physical reality – a much more complex and mystical picture than we are taught exists.

So, these are the layers of our spirit body that make us, in our Earth incarnation, complete people. We are not just physical beings, but complex individuals with an ability to connect with higher levels of reality. All living things, including animals and plants, are said to have etheric spirit bodies. The consensus seems to be that animals have a more complex spirit body than plants, but the spirit bodies of animals are simpler than those of humans. It is believed there are many more layers reaching up to Source or the Godhead.

The spirit body is multi-layered and connects with Source or the Godhead.

The energy in the food we eat powers our physical body, enabling it to function. Our spirit or subtle bodies also need replenishing and balancing to work efficiently. There is nourishment all around us in the form of universal energy, a life-force known as prana. This is fed into our subtle bodies via our chakras, which both take up and collect this life-force energy to sustain our subtle energetic systems. It is unlikely our physical bodies could exist without them because they act as gateways for the flow of energies which ensure our survival.

There is a universal life-force energy or prana which is essential for nourishing both our spirit bodies and our physical bodies.

Chakras and Healing
Chakras are centres of rotating energy, usually shown as a line running down the centre of the human body. The word 'chakra' literally means *wheel* in Sanskrit. All living things are supposed to have them. Human beings have at least seven main chakras and some people feel many auxiliary ones as well. Their existence and uses are well documented in alternative and healing circles. They are known in many cultures throughout the ancient world.

Our chakras are associated with different parts of the body and provide it with the energy it needs to properly function. Usually illustrated as coloured circles, our seven main chakras are generally accepted as follows: root (red), sacral (orange), solar plexus (yellow), heart (green), throat (blue), third eye (indigo), and crown (violet). The openness and therefore flow of energy through a specific chakra determines our state of health and sense of well-being. It is said that our chakras are connected to the body's main organs by energy rivers, or meridians, which flow through the physical body. It is also believed they are linked to the endocrine glands that secrete hormones to enable our organs to function.

Many people have developed the necessary skills which allow them to work on healing the physical body without the use of modern medicine. Some use the properties of herbs and natural remedies; others use the vibrational frequencies of homeopathy and essences. The properties of crystals, sound and colour can also be used for healing. Therapists practicing acupuncture, acupressure and shiatsu, for instance, work on the energy meridians running through the physical body. These ancient practices and healing methods are now beginning to be accepted and taken more seriously by some factions of the Western medical profession and, indeed, by increasing numbers of the general public. They are successful to varying degrees, even though conventional medicine is yet to recognise the physical mechanism, especially in pain control.

Reiki practitioners and spiritual healers, among other forms of alternative therapists who use non-touch forms of energy healing, use their skills to balance the chakras and heal the earth-bound layers of the spirit body. Any emotional, physical or psychological disturbance or imbalance held in the spirit body can cause problems with individual chakras. Many healers work using the universal life-force energy to repair any damage by first clearing any blockages, and then rebalancing the chakras. This allows free-flowing energy into the spirit body via the appropriate chakras for optimum health to be restored. Once rebalanced, the theory is that the spirit body can then tune-up its physical counterpart, enabling it to fight infection or put right any imbalances within its physiological systems. This can help both specific conditions and improve overall health and well-being. It should be mentioned, however, that imbalances within the chakras can be the result of stress, injury or trauma, also poor diet and lack of exercise, so a healthy lifestyle is paramount.

Some people claim they can see energy coming straight down though the crown chakra; others say it is more complicated. As previously stated, people perceive differently owing to the sensitivity and efficacy of their physical sense organs and level of spiritual insight or awareness. Speech is also vibrational energy and our thoughts can travel across great distances. So, we should take care not just in what we say, but also in what we are thinking!

Although the chakra system outlined here is generally accepted in mainstream spiritual and healing circles and, importantly, in complementary medicine, there are practitioners who cast doubt on its authenticity. There is currently no scientific proof of the existence of chakras, but there is substantial, long-standing anecdotal evidence, over thousands of years. Those who practice meditation also vouch for their validity. The sheer number of people successfully healing with the spirit body, using the universal pranic energy, indicates that it almost certainly exists. What's more, ancient (and some modern) wisdom traditions maintain that the spirit body is itself many layered and reaches up to the Godhead or Source.

Human beings have spirit or subtle bodies which enable the physical body to function properly. These are the layers practitioners work on to rebalance the chakras and facilitate healing.

The next important question I asked myself was, if this many-layered spirit body is present in all human beings, then is it part of what we refer to as our soul? If so, what does this mean for who and what, as human beings, we truly are?

The Soul

The straightforward definition of soul is the spiritual or immaterial part of a human being, regarded as immortal. This is what most of us have been taught in religious education. It is accepted by those who believe in such a thing. Where our soul is supposed to come from is another matter.

The conventional among us may say it is a gift from a creator God. More alternative thinkers often assume that although our souls appear to be part of what makes us an individual, they are fragments of something called Source. Thus, we are both individuals and part of a bigger collective. Our consciousness, life force and spirit bodies are all aspects of our soul that enable us to incarnate as humans. It is the spiritual part of a person which is not only believed to give us life, but to live forever. This, of course, adds another dimension.

The soul can be seen in two different ways. There is the soul aspect within each one of us that takes on the reflection of God, and through self-realisation we can reconnect with. The soul can also be related to our physical bodies – we talk of "the depths of our soul" when we feel empathy for others or experience beauty, for instance. Ancient wisdom tells us that only when our souls evolve from identifying with our physical bodies to the realisation we can reconnect with God or Source, can we reach a state of bliss. We then have the potential to become pure souls and have no need to return in physical form.

This is all unprovable, of course, and the definition one accepts depends on personal belief. As mentioned in Chapter 2, there is a huge amount of anecdotal evidence that human consciousness does go on in some form after physical death. Accepting this as fact is paramount to accepting there is an immortal part of each human being. The very existence of the soul gives us an important Super Clue.

The true nature of the human soul is a mystery that may be the key both to life after death and a multidimensional reality.

Whether our soul essence equates to the soul in a religious sense is questionable. In the *Bhagavad Gita* or 'Song of God', Krishna refers to the soul as the "indwelling Self". He tells his disciple, Arjuna,

"This Self is not born, nor does it perish. Self-existent, it continues its existence forever. It is birthless, eternal, changeless, and ever the same. The Self is not slain when the body dies."[4]

Some people believe that animals too have souls. They certainly have spirit bodies, but do they have that extra 'something' alongside their physical bodies? There is also, as reported, varying levels of conscious awareness, depending on the species. Some animals do seem to be able to communicate, at some level, with sensitive humans. As with us, the question is whether this is down to genetics or comes from elsewhere. There is, of course, much contention concerning this claim. Although, if you believe, as some do, that we may choose to reincarnate as animals, then presumably those animals will have souls. Also, some people feel they can communicate with lost loved pets, or that their animals, frequently dogs and horses, are communicating from the other side. Therefore, to exist beyond physical death, does this mean the animal must have a soul? This belief may, however, just be a factor in the make-up of their owner.

There seems to be an immortal part of humans, and possibly animals, that continues after physical death. Our spirit bodies, as well as our consciousness, may be fragments of what we call soul.

---

[4] Chapter 2, Verses 18 & 20. Cited by Swami Kriyananda in *The Essence of the Bhagavad Gita*, 2008.

I found that any research into these matters seems to lead back to something referred to as Source. My next question, then, was what does the term Source mean?

Source
The concept or existence of Source is outside of the sciences and beyond most people's imagination or understanding. While definitions vary, generally the term refers to a universal spirit, an intelligence or conscious will, responsible for all of creation. As human individuals, we may be fragments of Source.

Many people believe that our spirit bodies are connected to something called Source.

The manifestation of God's power is seen in the creation of the world in which we live. Could Source be the same as God or an intelligent designer? Absolutely, or it may be something completely different. As in the source of a river or stream, it is the place of beginning. By implication, Source is from where we originate, a part of the evolving intelligent mind and spirit. Some would say the creative force is universal as it transcends human nature; we are all part of one creative intelligence. Therefore, it could be humanity, collectively, and not just a creator God, responsible for our reality. If we associate the creative force with divine powers, and we do actually have the ability to co-create with the Divine, this makes the argument that we are responsible for our reality even more plausible.

What do the Super Clues embedded in the concept of our spirit bodies mean for our reality?

Summary
These aspects of reality, and their Super Clues, have moved beyond science into uncharted territories and unfathomable waters. They cannot, as yet, be substantiated, other than by the personal experiences of many people. One thing we can say for sure is that the human body is far more than a physical entity. The spirit body seems to have many layers, some used for maintaining or healing the physical body, which may themselves be connected to a source beyond our present understanding. There is also an immortal part of us, perhaps connected to the

Divine, which some say is an aspect of our human soul.

This study of the human spirit body, and its implications for animal and plant life, opens-up possibilities of a creation beyond that of a conventional God, or indeed of any intelligent designer. It gives us glimpses of an unknown and powerful big picture in which we, as individual souls and members of the human race, are fragments of a universal collective intelligence. We may even be co-creators of our own reality.

We are planted here on Earth during each incarnation and are, perhaps, not what we thought we were at all. With its many wonders and a multitude of possibilities, reality is, we can be assured, not just rooted in physical life on planet Earth.

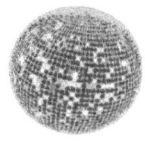

CHAPTER 5

Super Clues embedded in Life After Death

WE ARE NOT THE ONLY generation to have asked questions about the nature of our reality. Fundamental discussions and glimpses, of what in this book are called Super Clues, can be found in ancient writings and traditions throughout the ages. The ideas and theories of philosophers and scientists from hundreds, if not thousands of years ago, are teeming with debate about the nature of the universe which are still pertinent today.

The Greek philosopher Plato (c.429-c.348 BC) describes the material world as "a world of shadows". Plato's *Allegory of the Cave* was intended to teach his students that a shadow cannot exist without an object and a source of light – the image projected then becomes the reality. The Sun illuminates everything and makes it knowable, yet to Platonists the world of ideas is more real than matter, which is subject to the laws of time and space.

While this book does not venture into the realms of philosophy, the original blueprints for humanity came from the land of the gods – holograms perhaps? (see Chapter 1). There was certainly a belief in reincarnation in Plato's time. I am always in awe of how much the ancient thinkers understood, given they lacked today's modern scientific knowledge.

In this chapter, I also look at the realms of ghosts and spirits, which yield some important, if presently scientifically unprovable, Super Clues embedded within them. Many of these concepts about life after death show a much bigger picture of reality than is generally realised. They are rooted in ancient wisdom traditions, which modern science, in some areas, is beginning to catch up with. These include theories about karma and reincarnation, the Akashic records, and the concepts of good and evil – all of which, throughout history, have permeated our world of ideas about life beyond the physical.

The Soul Beyond Death
What really happens when we die and leave our physical body? This important question is a key to understanding our reality. As discussed in Chapter 2, there are many clues which seem to indicate that part of us, at least, does go in some form after physical death. The question is, when our soul consciousness leaves our physical body and passes over, where does it go? What Super Clues are hinted at in this transition? We could, of course, as some atheists believe, just expire into nothingness. I also discuss evidence that there is continuation, in some form, after physical death. The ancients certainly seemed to believe so.

Those of religious faith will generally assume some sort of passage to a Heaven (or Hell), leading to judgement by a creator God. Thus, the soul simply returns home to its maker. For atheists and those of non-conventional beliefs, there are alternative scenarios. As we have seen, science tells us we live in a multidimensional reality. There are almost certainly other dimensions and beings around us that we cannot see and may not even be aware exist. This includes the nature spirits and elementals (discussed in Chapter 9.).

In my research, I came across a dimension, acknowledged in some form by both ancient and modern thinkers, called the astral plane. This was believed by the ancients to be a plane or dimension that is crossed by the soul in its astral form or spirit body on its way to being born, and again after death. The astral plane is also allegedly the celestial realm where angels and other immaterial beings reside.

This plane may also be inhabited by unfriendly beings and those not so helpful to humanity, including disembodied spirits. Therefore, when connecting with the astral realms, do be careful to protect yourself from unwanted entities. Some of these may be souls who have passed over and, for whatever reason, been unable to move on. Spirit release is a specialised area of work and not for the feint hearted.

Another more modern view suggests the astral realm refers to the whole of spiritual existence. All those who leave their physical body return here to live non-physical lives. It is much larger than the physical universe, considered by some to be a place of peace and tranquillity. This seems to relate to the religious view of Heaven. Others are not so sure and believe the astral realm may also encompass Hell.

Heaven and Hell are thought to be outside of physical creation. In Islam, known as "The Barzakh, olam mithal" or intermediate world, where Jannah is the final abode or resting place – a garden usually referred to as Paradise. In Judaism, Heaven is the world of "Sepher Yetzirah". Yetzirah – the Book of Creation – is the third of four worlds in the Kabbalistic Tree of Life.

Other than testimony from those who believe they are in communication with the dead, and from individuals who have had near death experiences, there is no hard, scientific evidence for the existence of Heaven. Nevertheless, people hold varying personal views about what happens after physical death and there is plenty of anecdotal and empirical evidence, most of which involves passing over to other realms in some form. Alternative scenarios include:

- We return to our creator to be judged.
- Between lives we return to the astral plane to review our lives and rest, before being reborn.
- The soul returns to Source, from where it originated. An aspect of the soul then returns to a new physical body, with no memory of former lives.
- We find ourselves in a different level of creation, depending on how we have conducted ourselves in this lifetime.

Which of these, if any, is correct, only time will tell. The truth may, ultimately, be quite different. What seems to arise here is two versions of reincarnation: the soul either returns to Source or to the astral plane. The question is, are they the same or different realities? The general presumption of some sort of life after death does, however, give us concepts which are rich in Super Clues.

As our souls do seem to go somewhere between lifetimes, there must be a place, beyond visible creation, for them to go. The nature of this 'somewhere' is indefinable.

If we do, indeed, pass over and are reborn after physical death, our reality surely cannot encompass only that of the physical earth we perceive ourselves to be living on. The possibilities are endless. The astral plane, Heaven and Hell, other levels of creation and dimensions, or possibilities we have yet to consider, are all likely candidates.

Out of Body Experiences

Who and what we are is not just dependent on the functioning of our physical brain. The human brain is so complex that how it really works is not yet fully understood. Up until now, the main evidence for the brain and conscious mind being two systems rather than one (as discussed in Chapter 2), is substantial but anecdotal. Many people have tales to tell or have had near-death experiences, which confirm that a part of our consciousness can and does leave the physical body. Numerous accounts of those who have survived clinical death have been documented by Dr Raymond Moody, who first introduced the term "near death experience" (NDE) in his book *Life After Life* (2001).

Several phenomena seem to indicate not only does our spirit or consciousness separate from our physical body when we die, but that our consciousness can also leave our bodies while we are alive. People throughout the ages who have survived near death, or have technically died for a few moments or longer, report experiences such as floating out of their bodies, seeing a white light, finding themselves in a tunnel, hearing music or bells, and/or being greeted by deceased loved ones. They also offer proof that out of body experiences (OBEs) are possible by reporting seeing and hearing things which they could not have known or been aware of unless they had been there. In some cases, people have been able to tell doctors what was on the top of cupboards, for example, unseen from their hospital bed. Others, in severe pain or trauma, report floating out of their bodies for a short time. This has happened to a friend of mine.

Some people claim to be able to leave their physical bodies at will, often at night or in dreams. They can travel or fly around unseen, using the vehicle of their so-called astral body. There is also the recorded existence of communication with the physically dead. Could it be the astral body that people experience as ghosts or spirits?

As well as the vast topics of reincarnation and karma, which we will come on to shortly, these phenomena show us two things. Firstly, our soul consciousness does seem to go on in some form after physical death; mind and body being two systems, interacting and working together to make us human. Secondly, they may indicate the existence of a separable spirit body for each one of us.

Where our consciousness comes from or goes after physical death may be a clue to the ultimate big picture of reality.

Ghosts

Seen, recorded and talked about since civilisation began, even by people who claim not to believe in them, the existence of ghosts is the most written about of all 'otherworldly' phenomena, giving proof that we go on in some form. Ghosts are basically the energetic representations of people when they were alive – they appear to be once living people imprinted on our reality. People who have died after an accident or trauma, for example, may become stuck in a certain place from where they are unable to release themselves and move on. Those who have lost loved ones can sometimes see and communicate with their spirit on the Earthly plane. Others can see the deceased around them, and sometimes even smell their favourite perfume. Although some people do see the spirits of their dear departed after they have left their physical bodies, these, however, are not ghosts. I have myself seen dark shapes moving across a room.

There is a huge amount of empirical evidence: written, spoken, filmed and electronic voice recorded (EVP). Whatever their true nature, the very existence of ghosts gives us an indication of realms or dimensions beyond our physical world. Something else is surely going on.

Ghosts may be seen in different forms. There are:

- Those who walk about a place, attached to a certain location, often the victim of historical trauma. They are usually unaware of any changes to the buildings and will walk through walls, especially if new.
- Those who passively interact with people, seemingly fixed in the location they inhabited when alive, often their former home. Some do not realise they are dead.
- Those unhappy spirits that 'haunt' a location and are unable to move on to the next dimension.

Several types of ghosts, usually energetic representations of people who have died, seem to inhabit our reality.

Many psychic mediums report communicating with those who have passed over into spirit. They are also sometimes able to help those souls unable to move on. The Spiritualist Church is dedicated to proving that life goes on after death. Many open their doors to visiting mediums and offer spiritual healing. Although there may be some fakes or charlatans operating in this field, there is overwhelming, albeit anecdotal, evidence that communication beyond the grave is possible. It is through a resonant vibration that some people manage to do this.

Some people have an ability to see, and even converse with, spirits of the dead.

When ghosts or spirits occupy a house for any of the above reasons, it is said to be haunted. I happen to know someone who 'accidentally' bought such a house. The ordinary, unsuspecting family were plagued with strange manifestations and goings-on, even poltergeist activity.

Similar real-life stories are recorded by Christian Kyriacou in his book *The House Whisperer* (2014). He reveals that where we choose to live can support or hinder us on our soul journey, taking us into realms beyond our everyday reality.

Disembodied Spirits
As well as ghosts, there are other disembodied spirits reported to be in our reality with us.

- **Poltergeists**
  These are the "noisy ghosts". They are a type of ghost or spirit responsible for physical disturbances, such as loud noises, sudden bangs, and objects being moved or destroyed. Sometimes personal items and objects can be "spirited away" by "friendly" spirits, simply wanting your attention. Poltergeists are, however, allegedly capable of pinching, biting, hitting or harming anyone in their midst. They seem to have the ability to interact with the energies of disturbed humans to intensify their activities.

Some ghosts or poltergeists that interact with humanity may be acting with evil intent.

- **Demons**
  A demon is a supernatural and often malevolent spirit being. They are said to interfere with humanity, make you feel miserable and feed off the negative energy created. The emotion of fear also appears to attract those creatures hostile towards humanity. There is a belief in some religions that demons are agents or servants of

the devil and can cause demonic possession. Whether or not this is the case, in Western occult circles they do seem to be part of the negative, dark side of our reality. In some cultures, the underworld is where many demons reside. They could also be an alien species preying on humanity. They may, indeed, just be another species evolving alongside us, almost in contrast to the light energies of the angels (see Chapter 8).

Beings from the dark side that promote and feed off negative energies are with us in our reality.

There seems to be a universal battle between the forces of light and dark throughout the cosmos. What demons are and where they come from is unknown, but if they manifest or exist in another realm then there must be a place for them to come from. Some places seem to act as a conduit or portal for the more malevolent ghosts and spirits (or even the non-malevolent) to re-enter the Earth's atmosphere. Sometimes they appear to travel through portals and return to specific locations on the Earthly plane.

There may be portals to other dimensions or places in the cosmos from where malevolent spirits gain access to certain locations on Earth.

What does the existence of ghosts and spirits tell us about our reality? The very existence of ghosts and poltergeist activity provides some sort of proof that physical life on Earth is not the whole story. While some ghosts are trapped souls, and may or may not cause disturbance, there are other beings around that are good, bad or indifferent to us. The existence of malevolent spirits shows us that the concepts of good and evil, selflessness and selfishness, are an integral part of humanity.

If ghosts and poltergeists are coming from a different realm, then there must be another realm for them to come from.

Karma

Karma is sometimes thought of as getting what you deserve in life, according to your actions. It is often referred to in casual conversation, but what does it mean? Well, along with the concept of reincarnation, karma is a Buddhist principle. It may, of course, have been around for much longer. In some cultures, karma is one of the universal laws of wisdom, an aspect necessary to bring harmony and balance. The definition of individual karma seems to hinge on the concept of "you reap what you sow". Simply put, acts of unkindness or dishonesty will rebound on your life in a 'bad' way and result in suffering, while 'good' actions and intent are rewarded. Karma is seen in some ways as a kind of cosmic bank; behave with honesty and integrity, you will have a happy life, and vice versa. Our Earthly lives are intended to provide lessons and give us challenges. Who knows if these would be even harder, if we behaved badly?

The question is, why do seemingly good, kind people, who we may feel should have positive karma, still have difficult times or seemingly unhappy, undeserved things, happen to them? The answer, interestingly enough, is that karma does not just extend to this life, but also to past and future lives. Such individuals may be paying a price, working out karmic debts with people and situations from a previous lifetime. Therefore, to make sense of this and to accept the concept of karma, we have to believe in reincarnation – that we have lived before. The general view seems to be that our souls are seeking lessons acquired from many lifetimes here on Earth, and probably on other planets too, in order to further our spiritual development.

When we are beyond bodily incarnation, we make individual soul contracts to experience life in all its variations. This is so other levels of humanity can be experienced, and any karmic debt repaid. We play different roles in experiencing both sides of a situation, sometimes incarnating with other family members or friends. Some astrologers believe we even choose our time and place of birth, which determines our character and personality, strengths and weaknesses. The position of Saturn in our natal chart may signify our karma in this lifetime.

When we are born as a blank slate, so to speak, with no memory of our previous lifetimes, we forget our agreements at a deeper level and, while we have the free will to make decisions, will play out our lives reacting to what is thrown at us. Often, if we do not get the message, the same patterns will repeat themselves, even with the same people or soul group. This continues until we have accepted or forgiven our protagonist or come to terms with a situation and learnt the intended lesson. Only when we reach resolution is our karma balanced, and universal order restored.

Some people believe karma is just a myth. Of those who claim otherwise, there are differing views on where karma is stored: it may accumulate in our spirit body, it may be stored in the nervous system in the form of stress, or be recorded in what is called the Akashic records. David Russell offers another perspective on how karma functions:

> "Karma is the law of action and reaction. Every action, thought or word produces an effect that vibrates outwards into the cosmos. When that vibration hits an obstacle, it returns to its creator as karma. Karma is an essential and unavoidable element in every life. Most of us call it luck. Karma comes in many shades and colours. It ebbs and flows, sometimes good, sometimes not so good, and sometimes (the best times) seemingly not there at all. There is only one way to overcome karma, but it is not within the scope of most of us. Nor is there any way to penetrate its depths. Karma is an ineffable and apparently mysterious force within all our lives and the only thing to do is grin and bear it."[5]

In terms of what the mysterious force of karma means for our reality, it pre-supposes life after death and a cosmic plan of some kind. A belief in karma is therefore a belief in universal balance and the living of many lives. We are given choices each lifetime but have to pay our dues, with opportunities to resolve karma created both in the past and during our present lives. If it does exist, karma is a Super Clue to an ordered cosmos beyond physical existence.

The principle of karma opens-up the possibilities of many existences within a cosmic master plan for us all.

Reincarnation

Some faiths, especially in Eastern cultures, believe we live more than once and that our souls (or an aspect of our souls) reincarnate in different bodies in another place and time. In our discussion on karma, the assumption is that we have experienced many lifetimes to accumulate and work out any karmic debt. Implicit in the concept of reincarnation

---

[5] David Russell is quoted on my website, before his passing in 2016. http://www.marianmatthews.com/another-view-of-karma/.

is a presumption that a part of us, our soul essence, goes on in some form after physical death and enters a new physical body each lifetime – many bodies, one soul.

The theory holds that we may agree or choose to be reborn as any sex, colour, race or creed. We can live life from all angles, in different roles and positions of social standing, over time. We can experience every possible combination of circumstances, for good or bad, should we want to. There are, however, certain patterns which appear to recur for each individual. This assumes, on one level anyway, a degree of planning for our soul experiences. Some people believe we enter soul contracts with certain people before each incarnation. This may include those we have encountered before; even geographical locations can have a resonance from the past.

The principle is based on a person experiencing many lifetimes, during which they 'improve' or 'regress' as they accumulate or reduce negative karma, until they become a pure soul. They finally lose self-awareness and enter the boundless soup of the cosmic mind or go on to a higher plane when reincarnation is no longer necessary. The nature of the cosmic mind and what and where this higher plane may be is, at the time of writing, unknowable.

Could life indeed happen in this way? Well, there is presently no scientific proof, although reincarnation has been accepted as a reality in many religions and cultures for millennia. There is a lot of anecdotal evidence out there, as corroborated by Dr Brian Weiss in *Same Soul, Many Bodies* (2004). There are also books written by those who feel they have lived before. Josephine Sellers is a case in point, having returned to her former home in Dorset and met up, in sometimes challenging circumstances, with others who were once a part of her soul group. She tells her remarkable story in *The Return of Yesterday's People* (2002) and *Parallel Worlds* (2011). In other instances, children have gone back and identified people and places encountered previously, including things they could not possibly have known unless they had lived there. Many such cases are documented by Dr Jim Tucker in *Life Before Life* (2009), and by Dr Wayne Dyer and Dee Garnes in *Memories of Heaven* (2015). This substantial body of evidence cannot be lightly dismissed.

As noted, we are supposedly reborn with a clean slate, with no knowledge of our past lives, but sometimes there is a certain resonance, likened to bleed-through. Some people can access their past lives by meditation or with the help of an experienced psychotherapist. Perhaps the best-known accounts of past life regression are to be found in Dr Raymond Moody's *Life Before Life* (1991).

Although hard scientific evidence is not readily accessible, there is a substantial amount of oral and written reports from people who are able to recall past life memories and have even found historical evidence of their former lives. On occasions, memories just surface. To be aware of a past life can sometimes explain certain attitudes, skills or challenges faced in this current lifetime. Even one's irrational fears can be clues of past life trauma.

I have myself experienced past life memories surfacing from time to time. For instance, I know that I was once trapped under a boat in a cold northern sea and drowned. I can still feel the heavy beam on my chest and the fear and sadness when I could not escape. This may explain why sailing has always freaked me out, even though I am a good swimmer and not afraid of water!

Realising you have lived before and will likely live again certainly puts one's present life into perspective. It is therefore important to treat all around you well, as you never know, you may find yourself with the boot on the other foot, so to speak, next time. If aspects of this life don't suit you, then do the best you can and prepare to make changes in the next.

There is very strong evidence that life continues after physical death, that the physical plane is not all there is for human existence. This is a powerful Super Clue, which indicates a much bigger picture than life on Earth would generally lead us to believe. Importantly, if some part of us returns and we live many lifetimes, then there must be other dimensions beyond, to where we go between lives.

Memories of past lives indicate we may have lived many times before. The concept of same soul, different body, is a clue to a much bigger picture of existence than our present lives on the Earthly plane.

Assuming we are not all living an illusion, viewing the world in terms of the reflections cast by the analogy of a glitter ball, reality is certainly much bigger than our Earthly existence would appear to suggest. This does not, however, seem to be random; there does seem to be an element of design and record keeping involved. There is certainly said to be records kept specific to all mankind. This library is called the Akashic records.

The Akashic Records

In Sanskrit 'Akasha' means *aether* – air, sky, or space. The Akashic records are said to hold the energetic records of the past, present and potential futures of each individual soul, and some say our collective consciousness. They are believed be held in the non-physical plane of existence, outside of physical creation, known as the zero-point field. This knowledge is not just a New Age concept but ancient wisdom.

The Akashic records are generally thought of as being like a library, one book or record for each lifetime per soul. Some sensitive people can access these records, usually when in a meditative state. When envisaged to be in book form, this is what people will usually see. Others who work with the spirit body believe this information to be encoded in our individual auras.

Since both the development of computers and the growth in understanding of the underlying nature of the universe, specifically in the quantum and zero-point fields, the Akashic records have been likened to a cosmic computer database. It is worth noting, however, that although mainstream and conventional science may regard the concept of Akasha as non-scientific, those working with and/or studying quantum energies believe the energies of the zero-point field to underlie the fabric of space itself. In his book *Science and the Akashic Field* (2007), Ervin László introduces an integral theory of everything. He suggests the "quantum vacuum" is the fundamental life-force energy and information-carrying field flowing throughout the universe which informs our collective consciousness, not only in this lifetime but all universes past and present – collectively, the Metaverse.

Another view is that if we are all holograms anyway, Akasha may be the ultimate generating computer database. It is, nonetheless, a universal information field. So, these records are either in a big book in a cosmic library, or all around us bound in with all the energies of the universe, including our own consciousness. If either or both of these interpretations are true, this has massive implications for the nature of reality here on Earth. It would point to an organised structure beyond space-time, presenting a complex picture of creation way beyond our understanding.

There is a belief that the past, present and future of all souls is embedded in the structure of the universe via the Akashic records. This indicates a much bigger and pre-planned reality than we are usually aware of.

If the Akashic records are accepted to exist and we have lived before, this would indicate that physical life on Earth is not random after all. It may be designed and planned or predestined. Our lives may be embedded in the very structure of the universe itself, either within an energy field in this or another dimension, perhaps as part of a master plan.

What does the existence of the Akashic records mean for our reality? Simply, that if records exist and we have already lived many lifetimes, this must be for some kind of predestined purpose. Also, if our lives are designed or planned as part of the fundamental fabric of the universe, could there be a master controller somewhere beyond creation? Once again, this begs the question, who is doing the designing and planning? Could there, after all, be a God or an intelligent designer of some kind? Could we even be participants in a gigantic interactive computer video game?

If our lives are designed or planned as part of the fundamental fabric of the universe, there may be a master controller or higher power somewhere beyond creation.

There is another interesting reality twist. In 2016, Elon Musk, co-founder of PayPal and leader of Tesla and Space X (Space Exploration Technologies), came out with a statement that doubted our reality was the way we thought it was at all. He said the chance we are in "base reality" is small; it is almost fact we are living in a computer programmed reality, perhaps even developed by future civilisations. He argues that computer simulation is currently developing so fast, the difference between simulation and reality will soon be indistinguishable. There is a good chance that future civilisations have already developed this technology and put us into their simulation. Could this be possible? Could we be nothing but holograms in a computer simulated reality? We may even be living in a reality created by advanced alien beings or an artificial intelligence. According to a radio podcast,

"Musk is a firm believer in the hypothesis that a super intelligent artificial intelligence created the universe as we know it."[6]

---

[6] Radio Motherboard podcast, 2 June 2016. https://motherboard.vice.com /en_us/article/8q854v/elon-musk-simulated-universe-hypothesis

We may even be living in an alien computer-generated virtual reality, or a reality created by an artificial intelligence.

Good and Evil

Having been brought up with the concepts of good and evil as an accepted part of our Christian faith, this is of particular interest to me. The first question that arises is, do good and evil really exist or are they human constructs bolted on to our culture as means of control, especially in relation to the realms of Heaven and Hell? The second question is what do these concepts actually mean for our reality?

It is worth mentioning here that many people believe, on a higher level, the concepts of good and evil, heaven and hell, to be somewhat irrelevant. Some would say duality does not exist in terms of two independent principles but teaches us that every aspect of life is created from a balanced interaction of opposite and complementary forces. These concepts are, however, more difficult to pin down in practice than one might think. We all feel that we know what they are, until we try to define them.

- **Good**

The dictionary definition of 'good' goes on for half a page: "morally excellent, virtuous, not bad, positive rather than negative, etc. etc." None of these really define what the concept of good means in terms of behaviour. My sense is that how we conduct ourselves is what matters: doing our best to behave well and help others; causing no harm and acting where possible for the greater good in seeking to fulfil our own potential; not using others for our own purposes or to their disadvantage. This does not necessarily mean abiding with some of the narrow, artificial constructs imposed by state or church, but may mean following broadly religious principles, as guidelines.

Is the concept of 'good' then, an add-on, a power or choice given to us by a designer God? Perhaps it is a natural evolutionary development allowing us to live in relative harmony? It could be worth mentioning here that our distant ancestors lived in accordance with the good of the tribe. Acting from a place of ego or making decisions purely based on self is believed to be a relatively modern advancement. When psychology was established as a discipline separate from philosophy and biology, the study of conscious experience became one of the first topics studied in the early twentieth century.

The focus of this research began with observable behaviours before looking at an individual's level of self-awareness. Importantly, only with the development of language is a person able to describe in words an intention and what they are experiencing or feeling.

- **Evil**

    The definition of 'evil' is, as ever, a problem. We all feel that we know evil when we encounter it, but, as with 'good', pinning it down is more difficult. My dictionary defines evil as "profoundly immoral and malevolent". It also refers to bad character, working against the greater good, etc. Extreme selfishness can sometimes result in seemingly evil acts, especially when seeking one's own ends to the disregard of another's welfare. The kind of behaviour that causes harm or pain could be the result of random excessive selfish genes, or a psychopathic brain function with no apparent empathy circuits, allowing a person to put themselves at the centre of their own universe to the detriment of others. Malevolence does, however, indicate a desire to hurt others through one's own actions. Does lack of conscience or bad upbringing lead to seemingly evil acts? Could there be a force seeking to tempt people in this way for some kind of pleasure or personal gain, or is evil just a set of bad decisions? Obviously, the answer is influenced by one's individual beliefs and boundaries, giving us our own acceptable definition.

    In a Christian culture, we are brought up to believe that we are created by a 'good' God, and the existence of 'evil' is a conflicting force to be fought against at all costs. Evil had a leader called the Devil (or Satan), a fallen angel who defied God and fell from grace. Early Christians rejected the pagan gods who were believed to be evil spirits; they existed under many names to tempt you onto the wrong path. Failure to do good, or at least not be tempted by the darker side, resulted in retribution or even damnation. In Wicca tradition, people also use black magic, calling on the Devil to manifest their desires. The jury seems to be out on the existence of an actual perpetrator, the Devil, as the personification of evil.

I have also come across an interesting theory that if evil does exist as a force, it was purposely created by God to give us something to test ourselves against. Others believe both good and evil were created by an even higher power, way beyond our grasp. I personally have seen, or been on the wrong side of, intense selfishness, but fortunately have not come across what might be termed malevolent evil.

The question of whether we need evil comes back to the conundrum of why we are here? From the religious perspective, we may need evil to give us something to be tested against. To resist temptation allows us to reap our reward after death. For the spiritual but non-conventionally religious, this also applies. We need the dark, the light, and the shade, to aid our soul development. If we are not to learn and grow, why bother to incarnate? The corollary of this is that for us to need evil, we also need the perpetrators – those who are evil or bad. I have been told that some souls volunteer, before incarnation, to play this role.

If religious beliefs about the existence of good and evil are true, then beyond creation must exist a heaven, paradise or hell. Let us look at various beliefs about Heaven and Hell.

- **Heaven**
  While Christianity has a belief in an afterlife with the Divine, called Heaven, Judaism refers to the Garden of Eden. Christianity's roots in Judaism are reflected in the New Jerusalem, as described in the book of *Revelation*. The Christian concepts of heaven and hell are closely associated with religious ideas of salvation. It was the ancient Persians who gave us the word 'paradise'. The beliefs of Zoroastrianism that righteous souls approach the 'Bridge of Separation' before reaching paradise, are believed to have been adapted or adopted by the Jews, Christians and Muslims.

  In Eastern religions, instead of a heaven, souls are usually offered some kind of release from illusion and suffering. In Hinduism, it is through ignorance of Brahman that lives are acted out under illusion; this creates karma and causes individuals to participate in the cycle of death and rebirth (samsara). Buddhists are seeking Nirvana, to extinguish the flame of desire which keeps souls tied to the life-death-rebirth cycle; this is also the end of suffering.

- **Hell**
  There is little mention in the Hebrew Bible of a version of hell, except as a place known as Sheol where spirits of the dead reside. The Jews do, however, have Gehinnom, a place of eternal burning where unrepentant souls are cleansed of spiritual impurities before returning to the presence of God. In Judaism, the term Satan usually refers to

temptation or a difficulty to overcome rather than an actual being.

In Christianity, the idea that the Devil governs Hell may have evolved from the 14th-century poem, *The Divine Comedy*. When God threw the Devil and his demons out of Heaven, Dante portrays him as a grotesque, winged creature with three faces.[7] In the first part of this poem, 'Inferno', Dante depicts Hell as a place of burning. In Islam, the being who rebelled against God is known as Shaytan. When the Quran refers to fire, it usually means a place of torment.

Eastern religions have a very different view of the afterlife. However, like Jesus, the Buddha also resisted the demon when Mara tempted him away from the path of enlightenment. The *Tibetan Book of the Dead* mostly deals with the state between life and death, the Bardo where the soul exists before reaching Nirvana or rebirth.[8] Samsara in Hinduism, the endless cycle of death and rebirth, is the result of our ignorance of the ultimate reality of the universe. It is believed to determine the nature of one's rebirth and the caste one is born into.

There are other views on the nature of good and evil, Heaven and Hell:

- There are heavenly forces for good or light led by God, and those for evil or darkness with the Devil at the head.
- Evil or a place of hell does not exist as a force but is man-made.
- Evil is negativity to the point of extreme malice.
- There is no Devil as such, only bad human selfish genes.
- Evil is not a force, but a will or drive, not to be ignored.
- What we believe in exists, so to believe in evil or hell means it can exist.
- We manifest our own reality; therefore, all acts are our own creation – we create our own heaven on earth.
- On a higher level, the duality of good and evil, Heaven and Hell, is irrelevant.

---

[7] Dante Alighieri, *The Divine Comedy: The Vision of Paradise, Purgatory and Hell*. http://www.gutenberg.org/files/ 8800/8800-h/8800-h.htm

[8] *The Tibetan Book of the Dead*, compiled by Padma Sambhava, translated by Robert Thurman, 1993.

The forces of good and evil permeate our reality. The extent to which these concepts affect our Earthly incarnation may depend on what we believe and whether we consider Heaven and Hell exist.

We can see there are many interpretations of good and evil, and what their existence or otherwise means for our reality. As with the religious concepts of Heaven and Hell, this may indicate a deliberate design by an overarching God to test our souls, or to manifest conflict and control. There is also a view that we move up and down the levels, which range from the lowest, dark, dismal and hellish, to the higher realms filled with Divine light and goodness. The duality of good and evil may, however, be opposite sides of the same coin, as it were. Equally, they may be a part of a master plan. It is worth mentioning that if good and evil exist as external forces, then the conflict between darkness and light may be an eternal cosmos-wide battle.

If the concepts of good versus evil exist, then this indicates a more deliberate, planned design and an element of control and battling forces, by an intelligence or overarching God.

What do the Super Clues embedded in life after death say about our reality?

Summary

The mysteries of life after death, the concepts of karma, reincarnation, and the Akashic records, are topics which have been discussed and pondered over for millennia. They are still valid today, in our search for Super Clues to reality.

Although there is some work being done in near death situations, which seems to indicate that our soul consciousness continues after death and does go on in some form, where it goes and whether it returns, is unknown. The truth, as in all these matters, is not generally provable by today's science. All we have to go on are the insights of those with personal experience of past lives, near death experiences, communication with the spirit world, and the ancient knowledge that we incarnate throughout many lifetimes.

The ability to manifest or alter our reality would seem to suggest we are masters of our own universe. We may not, however, be in control of our own individual realities to the extent we think we are. Similarly, if good and evil exist in the greater scheme of things, and are not simply human constructs reflected in the realms of Heaven and Hell, are we just pawns in a vast game or unwittingly part of some universal battle? These concepts also have implications for what happens to our souls after death.

These Super Clues give us indications of a more multi-layered cosmos, with many more dimensions and forces existing in our reality than we are generally aware of or think much about.

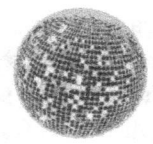

CHAPTER 6

Super Clues embedded in Universal and Earth Energies

AS EXPLAINED IN PREVIOUS CHAPTERS, we are essentially no more than vibrating energy in a cosmic energy matrix. Throughout the universe are the immense forces that hold our universe together. These include gravity, electromagnetism, and the atomic forces bound up in the structure of matter. There are also the quantum forces holding all of this together in our matrix grid to create our reality. Thus, the Earth is a seething mass of interacting energies.

There are clues that suggest these energies are part of a bigger cosmic system, which seems to indicate an element of deliberation. Some of these energies we use to ensure our survival, and some we can use to enhance our spiritual powers and for healing. Some are scientifically measurable, while others are not. There are those which are recorded and known, and those we interact with giving us glimpses or clues to a greater reality beyond what science tells us.

In this chapter, I will explore the Earth's magnetic and gravitational forces and how they interact in our solar system. We will also be looking at alternative cosmological theories including planetary grids, energy vortexes, and other kinds of earth energies including ley lines and dowsing. The Gaia hypothesis considers the earth as a sentient being. All give us Super Clues to a greater reality.

Solar and Lunar Energies

The heliosphere is the immense magnetic bubble emanating from the Sun that contains our solar system, solar wind, and the entire solar magnetic field. It protects us from the cosmic rays and dust beyond the solar system. The heliosphere is known to extend well beyond the orbit of Pluto.

SUPER CLUES EMBEDDED IN UNIVERSAL AND EARTH ENERGIES

- **Sun energy**
  The solar energies radiating from the Sun power our biosphere and enable the Earth to sustain life. The Sun is our main source of energy, providing heat and light. Sunlight heats the surface of the Earth and enables photosynthesis, which most of our plants use to grow and reproduce, kick-starting the food chain. Green plants are autotrophic and can obtain their food and energy needs from solar energy, some minerals, and just a few inorganic materials including carbon dioxide and water. Those which do not directly photosynthesize either leech onto other plants that do or get their nutrients from decaying matter in the soil. Thus, they are indirectly using solar power. Humans and other animals harvest the plants or eat the animals that have eaten the plants. The Sun gives us sustainable life and an ongoing food chain.

- **Moon energy**
  Our Moon is a round body, a natural satellite, with a radius of 1.737 km as opposed to Earth's 6.371 km, so about 27% of its size, but enough for its energies to impact the Earth. The Moon is believed to have once been part of our Earth, but to have broken away at some time in the remote past. There are several theories on how it was formed, and how it came to be where it is. The best contender seems to be that it happened billions of years ago when the Earth was still in a molten state and collided with another body in our solar system. The debris from this impact bound together to create our Moon, which was apparently at one time much closer to Earth.

  The Moon is held in place by the gravitation pull of the Earth, which it orbits every 27.3 days. It always has the same side facing us and appears to wax and wane in a 28-day cycle. The intensity of its brightness varies throughout the lunar cycle as the amount of reflected sunlight changes, depending on its relative position to the Sun in its orbit. Moonlight is a useful source for those without artificial light at night. The lunar calendar of 13 'moonths' can also be used effectively for planting and maximising natural growth.

  The energies of the Moon directly affect us in several other ways. Female menstrual cycles (usually 28 days) are linked to lunar cycles. There is an old established theory that the energies of the Full Moon can make people insane. Lunacy comes from the Latin word *Luna*, meaning Moon. In the same way as the Moon affects the tides, mood swings are also thought to be the result of the gravitational pull on our bodily fluids, affecting the brain.

Astrologers believe that the energies of all planetary bodies, and their relative positions in the heavens at our moment of birth, affect all of us throughout our lives. In Western astrology, the position of the Sun when we are born is said to represent our will and conscious mind. The Moon represents the characteristics of our subconscious impulses. Moon energies are the basis for a complete system of Chinese astrology, based on animal signs. In my first book, *Aspects of Reality*, I describe astrology as a "Satnav in the sky" which can help us navigate our way through life on planet Earth.

- **The tides**
Just as the Earth's gravity holds the Moon in place, the Moon's gravity pulls on the Earth. The most obvious visible result is seen in our tides as the vast areas of water on the surface of the Earth, the oceans, are pulled towards it. These moon-driven tides contribute to making life on Earth viable.

    Here are just some of the ways in which the tides are useful to mankind:

    - The rise and fall of the tides help to remove pollutants and circulate nutrients, which nourish the sea plants and fish that in turn help to feed us.
    - Ocean tidal flows help to transport heat from the equator to the poles. This helps to moderate extremes in temperature and make many places on Earth habitable.
    - Some theories of our origins suggest that life began in the sea; the tides helped to spread both primitive cells and the nutrients they need to develop early life.

So, the heliosphere protects us from space debris, the Sun nourishes and sustains life on Earth, and the Moon lights our night skies and drives our tides. They all work together in unison to make the Earth habitable and sustain life. Could this just be an accident or deliberate design? If so by whom? Personally, I feel that life on our planet is just too intricate, with too many factors needing to work together to allow us to survive and thrive, both on a human and planetary level, for it to be a coincidence.

SUPER CLUES EMBEDDED IN UNIVERSAL AND EARTH ENERGIES

The combination of heavenly bodies which allow life on Earth to survive and thrive may be a happy accident or the result of deliberate design.

The Earth's Magnetic Field

There are two major Earth grids which exist throughout the planet: magnetic and gravitational. There are also other grids such as the Hartmann and Curry grids which are oriented to the earth's magnetic fields, although their existence is argued over by scientists. These give us yet more clues to a bigger, cosmic picture.

The Earth's magnetic field, which extends from the interior out into space, is called our magnetosphere. It protects us from the charged particles of the solar wind, emanating from the Sun. In fact, the Earth can be looked upon as a giant magnet anchored by north and south poles. At the centre of the Earth is its core, made up of two parts: a solid, inner core of iron, surrounded by a liquid, outer core of nickel-iron alloy. It is from here that our planet's magnetic field is believed to be generated. This flow of liquid iron creates electric currents, which in turn produce the magnetic fields that generate electromagnetic energy. As they are affected by the underlying rocks, strata and cracks in the Earth's crust, these fields are not constant and naturally fluctuate. This means that the poles switch position periodically. We are said to be overdue for a pole reversal in our present Age of Aquarius.

Many aspects of our lives depend on the Earth's magnetic field, from the electrical grids that power our homes and appliances, to computers and satellites. The Earth's magnetic field is very weak. It ranges from approximately 30 to 60 millionths of one tesla, compared, for example, to magnetic resonance imaging (MRI) which uses magnetic fields in the range of 1.5 to 3.0 tesla. It does, nevertheless, protect us from any negative effects of the charged particles emanating from the solar wind. While scientists are not exactly sure how they do it, research using a photochemical compass suggests that about fifty animal species, including migrating birds, turtles and other mammals, use the Earth's magnetic field, along with light, to navigate.

Solar flares have been recorded since 1749. These are caused by sunspot activity which create stronger or weaker cosmic rays, owing to the ejection of plasmas and particles into outer space. When they penetrate the Earth's upper atmosphere (ionosphere) and these particles hit our magnetosphere, the Aurora or Northern Lights create a stunning light show in the sky.

There are two radiation fields surrounding the Earth. These are the Van Allen belts, concentric-shaped zones of highly charged particles which are trapped at high altitudes in the magnetic field of the Earth. Named after the American physicist who discovered them in 1958, the solar wind is responsible for these doughnut-shaped fields contained within Earth's magnetosphere, which change according to geomagnetic conditions. Both belts are separated from each other by an 'empty' region, which is wider during low geomagnetic activity.

Life on Earth is protected from the damaging solar wind by the Earth's magnetic field.

Gravitational Forces
The force of gravity keeps all planetary bodies in orbit around the Sun. It keeps us on the surface of the Earth and stops us spinning-off into space. Physics tells us there is undoubtedly an energetic gravity grid flowing through the Earth. We deal with gravity every day and tend to feel that we know what it is. Strangely, its exact nature is more contentious than one might think. The word 'gravity' comes from the Latin *gravitas*, meaning 'weight', and from *gravis* meaning 'heavy', defined in the dictionary as:

"the force that attracts a body towards the centre of the earth, or towards any other physical body having mass".

We know the Sun and the Moon keep us on the surface of the Earth. So, is it the attraction between the celestial bodies that holds the Earth in position around the Sun, whilst keeping the Moon orbiting around our Earth? If so, why is our gravitational field relatively weak? We should, in theory, not be able to lift our feet off the ground!

It has been suggested that gravity is just another form of magnetism. This cannot be quite true, as magnetism repels as well as attracts and gravity does not appear to do so. As with our magnetic field, gravity is not constant throughout the Earth. Maybe it's down to a force between masses? Or is it, as Albert Einstein suggests, an effect of the warping of space and time in the presence of mass?

Let us look at some of the various gravitational theories, along with more recent findings from space explorations.

- **Einstein's view of gravity**
  In his pioneering work, *Relativity: The Special and the General Theory* (1916), Einstein determined that massive objects cause a distortion in space-time, which is felt on Earth as gravity. He described gravity as the motion of objects following curved lines in space – or rather space-time.[9] The curved lines are caused by the presence of a mass; the Earth or the Sun, for example. Therefore, this mass causes space-time to curve. Objects moving near a mass roll towards it, much as a ball would roll towards a person sitting on a trampoline.

- **Space-time**
  Einstein's theory of relativity becomes more complicated because as the Earth spins it should cause a twist in the fabric of space-time. NASA's Gravity Probe B (GP-B) confirms this prediction. So, gravity is not just a fixed energy, flowing through the Earth with all the other planetary bodies in our energy matrix, pulling masses together. Gravity also affects time, and time depends on gravity. A clock will tick slower at higher gravity than at lower gravity. As Einstein showed, time is relative, and a time dilation is always experienced by the observer (see Chapter 3).

- **Uneven gravity**
  Long-term mapping of the Earth threw up some interesting data about variations in gravity. In the GRACE mission, a collaboration between NASA and the German Aerospace industry (2003-2017), twin satellites mapped the Earth's gravitational field and charted its anomalies. Scientists looking at the quantity and movement of water found not only that the Earth is not perfectly round, flatter at the poles and fatter at the equator, but the strength of gravity varies enormously over the surface of the Earth. A shaded colour chart was produced, from all compiled data, which showed that gravity was found to be less strong in the oceans, stronger towards land masses, and at its strongest mostly in high mountain ranges.

- **Variable gravity**
  Notwithstanding the major gravitational differences as shown by GRACE, there are other variations. Gravity is not an even energy

---

[9] See also http://www.einstein-online.info/elementary/cosmology.html

flow, but changes across the surface of the Earth and throughout its atmosphere. This is owing to the effects of several factors:

- Gravity varies with latitude. We weigh about 0.5% more at the poles than on the equator.
- Gravity changes with altitude. The further away from the centre of the Earth, the less pull it has on the human body so generally we weigh less at altitude.
- Gravity varies with geology. Differences in local geology create very small changes in gravity (< 0.01%). The density of rock can have a slight effect on the gravitational force, which is greater in mountains where it exerts more force.

- **The Electric Universe**
  Modern science is questioning Einstein's theory of relativity and discussing what is termed 'The Electric Universe'. This controversial theory, which has various models based on recognising that natural electricity is generated by cosmic plasma (ionized gases), postulates that gravity is simply another manifestation of electromagnetism. It suggests reality can be better explained by electricity and magnetism than by gravity alone. In December 2019, the first scientific papers were published in *Space News* on the Solar Probe mission's earliest findings, using data from NASA's Magnetospheric Multiscale Spacecraft. The authors say the most intriguing discovery to date that the probe has detected is the "unexpected" changes in the Sun's magnetic field.[10]

Leys
A network of invisible lines of energy criss-cross our landscape, known as leys or energy lines. In Australia they are referred to as song lines, and as dragon lines in Eastern cultures. What causes them is a matter for debate. Some people believe that ley lines are caused by gravitational and electromagnetic energies that permeate through cracks in the Earth's mantle, or by boundaries in the tectonic plates. Others believe they are part of a fundamental crystalline grid construction of the Earth.

---

[10] Douglass, Scott, 'Donald Scott: Parker Solar Probe and the Electric Sun', *Space News*, December 14, 2019. https://www.thunderbolts.info/wp/2019/12/14/donald-scott-parker-solar-probe-and-the-electric-sun-space-news/

SUPER CLUES EMBEDDED IN UNIVERSAL AND EARTH ENERGIES

The Earth itself has a charge travelling through it, known as a telluric current. This is an electrical current which moves underground and through the sea. Telluric currents can result from natural causes, electromagnetic energies or human activity. They are very low frequency, travel over large areas, at or near the surface of the Earth, and may interact in complex patterns. These energies and the effects they have on surface phenomena are detectable, by some people at least, through dowsing. They often reveal themselves in the positioning of so-called leys, which are surrounded by contention and mystery.

Whilst not visible to the naked eye, the giveaway seems to be the ancient monuments, stone circles, burial mounds, holy wells and old churches aligned along them, usually in a straight line. In his classic book *The Old Straight Track* (1925), Alfred Watkins points out that leys are spirit paths, also associated with the living who used these ancient tracks to travel, stones marking the way. Many churches were built on the sites of pre-existing temples, along lines of sight recognised by our prehistoric ancestors. They include natural landscape features such as prominent hills and rock formations.

The best-known and longest alignment across southern England, known as the Michael line, runs from St Michael's Mount in Cornwall, through Glastonbury, Avebury and along the Icknield Way in the east, before going out to sea at Hopton in Norfolk. Dowsed and documented by Hamish Miller and Paul Broadhurst in *The Sun and The Serpent* (1987), it is not just a straight ley but has meandering male (Michael) and female (Mary) energy currents that form node points or power centres where they intersect. The significance of prehistoric sites that have survived along this 240-mile line, including in the west many old churches dedicated to St Michael, guardian of high places, suggests that our ancestors had geomantic knowledge which has since been lost.

The subtle energies emitted by leys seem to enhance spiritual energies, especially at sacred sites which are often aligned to sunrise at summer or winter solstice. The question is, were mounds, standing stones and churches located on these sites of known special energies for religious or secular purposes? Or did their existence subsequently create an energetic shift in the landscape energies, which may have been further enhanced at these locations?

Our ancestors built their sacred sites with knowledge of the stars and earth energies to enhance the land and to ensure their own wellbeing.

Dowsing

Dowsing is a traditional way of locating earth energies and many other materials. Using rods, a pendulum, or one's own body to feel the energies, subtle changes in frequency at or near the surface of the Earth can be detected. Dowsers today use metal rods but traditionally people used twigs, often hazel. Dowsing is used to find leys, water, electric cables, minerals, in fact anything that sends out an energy signal at a given location. It can even be used to find lost keys. When walking over any anomaly or identifying the chosen object, the rods will move apart, together or dip. A dear friend of mine began a lifetime's fascination with dowsing when he saw a Water Board official locate a water leak accurately with two old rusty strips of metal!

Energy lines flowing through the earth can also be identified in this way. People sometimes report feeling a different vibration or frequency in certain areas of the landscape, which they are then able to follow. Enthusiasts chart these leys on a map or explore those already outlined by others. These too can be measured; energy lines are not static and are often found to have a wider bandwidth at full moon. Archaeologists also use dowsing to identify boundary walls or remains of ancient sites, for example, and to understand more about the knowledge and skills of our forefathers.

The scientific explanation for what happens is something called ideomotor movements; muscle contractions caused by subconscious mental activity which can make anything held in the hands move. It looks and feels as if the movements are involuntary. Similar phenomenon has been shown to lie behind movements of objects on a Ouija or spirit board. It is, however, important to formulate one's questions clearly when dowsing and to always ask permission at the outset.

Dowsing can help us locate leys and earth energies and to identify water and other resources. Anyone can do it!

I was once shown a high energy spot by a family member. It was in an open field and, as I walked across the grass, my pendulum began to swing violently. More interestingly, I could feel the energy flowing up through my body. When a negative energy spot was pointed out, a feeling of great unease came over me as I stood on it. Strangely, the photos taken on that day would not upload properly.

On a purely physical level, we routinely interact with the earth when we walk, dance and sing – all these activities have a beneficial effect on the earth and the spirit of the land itself as we exchange our consciousness. When one of my dowsing friends tested the ground before and after Morris dancing took place, heightened energies were definitely detectable, but faded after a short time.

Pilgrimage can also help with an awareness of the sacred nature of our environment. As Peter Dawkins, elder of *The Gatekeeper Trust*, an educational charity devoted to personal and planetary healing through pilgrimage, tells us,

> "The Earth can be turned into light through pilgrimage, and there is a science associated with how this is done. Love is an energy, and all things including ourselves are built out of energy… If our energy is loving, then we affect other energy fields that we touch or move in with that love."[11]

Energy Vortexes
As the Earth spins, its rotation affects the energies that flow through the Earth. On the surface this is seen as changes in weather patterns and ocean currents. It also creates storms and whirlpools. The Coriolis effect is a force that acts perpendicular to the direction of motion and to the axis of rotation. This explains why water going down the plug hole swirls in a clockwise direction in the southern hemisphere and counter-clockwise in the northern hemisphere.

Every rock, liquid or strata allows the gravitational forces to travel differently through them, depending on the nature of the structure and mineral content. Because the Earth is constantly rotating, these unbalanced currents can form unseen vortexes. There are reported to be energy vortexes in many places all over the world, both large and small scale. Sedona in Arizona is particularly well-known for them.

Chris Harris, a dowser, has an interesting theory about so-called fairy rings being an indication of variable gravity vortexes, which can cause other distortions. He explains,

> "If planet Earth had a uniformly dense core (like a marble), the density of space generated by its mass would also be uniformly

---

[11] Peter Dawkins, 'How does pilgrimage help the Earth?' The Gatekeeper Trust. www.gatekeeper.org.uk

less the further out it is measured. But below the surface the Earth has many areas of varying density such as rock and lava masses. This gives rise to such features such as fairy rings that appear on the surface. The surface of the Earth is encased in an ocean of gravity that is susceptible to ripples, waves and whirlpools caused by variations in the density of the mass below the surface.

"I'm sure you will be familiar with these circular patterns seen in the grass of urban parks or even in the open countryside. The official explanation of these rings is that they are a fungus … The grass seems to grow at a different rate around the circumference of the ring; the rate is an accelerated or slower rate than the rest of surrounding grasses. Also mushrooms around the edge of the circle seem to grow uncannily well. The circles vary from just a few inches to many metres across and are found all over the world. I think that the gravity field that extends out from the centre of the Earth's mass (generally uniform) does have vortices in it coinciding with variations in the mass of strata below the Earth. Therefore, to view an open field is like looking at the surface of a lake upon which the gravity vortices like whirlpools can be seen via circles in the grass.

"… Gravity whirlpools give rise to fractionally faster and slower rates of time-flow leading to a visual difference seen in the age of growth of grass around the ring. I think ancient peoples knew of this effect and erected shrines (like Stonehenge) over the rings to amplify the effect. I might add, that whilst I am not an expert dowser, all the rings I test with my rods show a marked deflection when used over the rings. My local urban park is full of such features. Also, I think it is possible that two or more whirlpools could be connected by a straight line, as areas of identical spatial density (gravity) bleed into each other. Such straight-line connections would be an explanation for some ley lines."[12]

Another experienced dowser, Judith Lock, connects leys with interplanetary grids. She comments,

---

[12] Chris Harris, personal communication, 2018.
https://chrisharris.ucoz.com/index/dowsing/0-126

"Some leys appear to have an astronomical link, as do stone circles. These usually, but not always, are linked to the particular circumpolar north star at the time of making. Due to the precession of the equinoxes, others are linked to whatever was the spring house at the time. The Earth resonates to the Schumann resonance; we can't hear it, as it is just below our hearing range. It is argued that prehistoric man could hear it, and certainly some birds and animals can. Other planets resonate at different frequencies. It is said there is a symphony in the sky, but we don't hear it. The planets interact through vibration, so possibly the World grid and energy lines interact."[13]

As we have seen, the land naturally embodies energy lines of many kinds. If we build over vortex points, it will either activate or block the earth energy at that spot and create fragmentation, which can lead to what is termed geopathic stress. Together with disturbance of the water table that causes 'black streams', this can be harmful to health and lead to illness and dis-ease. These unbalanced energies can be identified through dowsing, which can also be used to help rebalance the land. Geopathic stress is a complex subject and can also result when the electromagnetic fields of the Earth are disrupted owing to mobile phones, tetra masts and other frequencies that negatively affect the Schumann resonance of the Earth and human brainwave patterns, depleting our body's own natural energy fields. The introduction of 5G (fifth-generation cellular wireless) using extremely high frequencies is of concern to many and could have harmful long-term consequences.

Planetary Grids
It has been known for thousands of years that energy lines criss-cross the world creating a global grid network. Native Americans used them to construct their medicine wheels and to contact the ancestors and spirit guides. Earth grids were used in the siting of the Great Pyramids at Giza, the megaliths at Sarawak (Borneo), Easter Island, and many other sites throughout the world.

Many sacred places can be found on major grid intersections, which may explain earth faults and magnetic anomalies. As Richard Lefors

---

[13] Judith Lock, dowser, 2018.
  http://healthwellbeing.focusonuk.co.uk/using-dowsing-holistically/

Clark, in his chapter on 'Diamagnetic Gravity Vortexes' in *Anti-Gravity and the World Grid* (1987), states,

"Pyramids and ley lines are on the power transfer lines of natural Earth gravity."

He goes on to explain that the grid makes use of geometrical flow lines of gravity within the structure of the Earth itself. This gives rise to the question, has this entire web of energy been used in the past to power ancient civilisations and could it be used in a similar way today?

Owing to the fundamental atomic structure of all physical creation and the laws of quantum mechanics, the Earth is an interactive energy system, part of a cosmos-wide interlinked net.

Some interesting alternative grid theories have been seriously considered by researchers over the years. While not always in line with modern scientific thinking, here are just two of the most prominent theories.

- **Crystalline planet**
  This theory is one of the Earth as a large crystal. Socrates, over 2,000 years ago, alluded to the idea that the Earth viewed from above would look rather like ball made up of 12 pieces of skin sewn together. Russian scientists in the 1960s published a paper, 'Is the Earth a Large Crystal?', suggesting that a matrix of cosmic energy was built into the Earth at the time it was formed, traces of which can still be detected today.

  In an article of the same title, in *New Age Journal*, May 1975, Christopher Bird explains this comprised 12 pentagonal slabs – a dodecahedron – overlaid with 20 equilateral triangles.[14] Other researchers working along the same lines have postulated slightly different values to the grid. Nichols R. Nelson, in his book *Paradox* (1980), saw the world as an Icosahedron (20 faces). In 1984, William Becker and Bethe Hagens demonstrated how the icosahedron and dodecahedron, a Rhombic Triacontahedron (30 face polyhedron), assimilated into the global grid, which they called 'Unified Vector

---

[14] See also Christopher Bird's website: http://vortexmaps.com/chris-bird.php

Geometry'.[15] Richard Buckminster Fuller, known for his geodesic domes in the late 1940s, together with other researchers who had their own theories, worked extensively on this quest to understand the geometry of the Earth.

A strange 'crystal', 4.5 billion years old, was reported in 2012 to have been found in the mountains of Russia, older than any rock found on the Earth's surface. Analysis suggests this was a fragment from a meteorite at the time of the formation of the solar system, before Earth even existed.

The Earth may have had a crystalline past, creating a matrix of inbuilt cosmic energy when it was formed.

- **Anti-gravity and the world grid**
  As already established, gravity and magnetic forces do not travel though the Earth evenly. The Earth is like a giant magnet and, as with any magnet, there is some weakening when the forces move away from the poles, towards the equator. This is due to something called a Bloch Wall, caused by the layer at the changeover between positive and negative magnetic fields. The charge is much weaker in the centre as the polarity switches. This means that at certain points on the Earth, the magnetic flow is very weak or even non-existent and leads to a phenomenon called diamagnetism. This is essentially a magnetic neutral zone existing between a north and south magnetic field. The Bloch Wall anomalies were, apparently, much further north some 10-15,000 years ago.

  There are many magnetic flow reversal points on the Earth marked by grid points. Lefors Clark in his previously mentioned chapter 'Diamagnetic Gravity Vortexes', suggests diamagnetism is at the root of anti-gravity, which is the force that enables levitation. He writes,

  > "Diamagnetism operates at 90% from magnetic, but in three directions, and not flat and two-dimensional as [is] usually drawn. If magnetism flows in the plane of the Earth's surface,

---

[15] Cited by David Wilcock, 'Becker/Hagens – The Global Grid Solution'. https://www.bibliotecapleyades.net/mapas_ocultotierra/esp_mapa_ocultotierra_15.htm

then diamagnetism flows straight up. Straight up is the direction that we call levitation or anti-gravity."

Ivan T. Sanderson asserts in *Invisible Residents: The Reality of Underwater UFOs* (2005), that there are twelve places around the globe situated on lines of latitude where the Earth's magnetic field gives rise to anti-gravity. Referred to as 'vile vortices', these are places where ships and planes have been known to vanish, such as the Bermuda Triangle. These vortex locations are allegedly known about but kept secret (as was the alleged Nazi base in Antarctica), and even fought over by scientists and military establishments. It is thought ancient civilizations had some knowledge of them, and that the energies at these locations can be harnessed and manipulated. They are portals into other dimensions. If you wander into one, you may find yourself living in a parallel dimension!

Other dimensions and the greater cosmos may be accessible via certain anti-gravity locations on the Earth.

Gaia

In Greek mythology, Gaia was the personification of the mother goddess who presided over the Earth. Although expanded and restated by James Lovelock in his books and lectures, the Gaia hypothesis is not an entirely new concept. In the 18th century, biologists and scientists were exploring and theorising about the Earth as a sentient being. For example, in 1785, James Hutton, considered to be the founder of modern geology, presented his *Theory of the Earth* to the Royal Society of Edinburgh. He believed the Church of England to be wrong and insisted the universe was far older than the accepted 6,000 years, declaring, time has "no vestige of a beginning, no prospect of an end."[16]

In his book *Gaia: A New Look at Life on Earth* (1979), Lovelock gives evidence for the interconnectedness and self-regulating factors that enable life on Earth to survive and even flourish. In 1974, a contemporary of Lovelock, Lewis Thomas, had already published *The Lives of*

---

[16] Cited by Morrice McCrae, College Historian, RCPE, 'James Hutton's Theory of the Earth…, 1785', *Journal of the Royal College of Physicians*, Edinburgh, 2012; 42:87–9. https://www.rcpe.ac.uk/sites/default/files/exlibris_2.pdf

a Cell, on the interconnectedness of nature and all living things.[17] Lovelock's later *The Revenge of Gaia* (2006), looks at the effect human activity is having on the planet. He defined Gaia as,

> "... a complex entity involving the Earth's biosphere, atmosphere, oceans, and soil; the totality constituting a feedback or cybernetic system which seeks an optimal physical and chemical environment for life."

It was Rachel Carson's *Silent Spring*, in 1962, that exposed the destruction of nature through the widespread use of pesticides and first caught many people's attention to the need to look after our planet. Was the Earth indeed the Great Mother nourishing her human children or was she about to abandon them? It was time to wake-up to our responsibilities in looking after the nature of our reality, the reality of Nature.

The energies of Gaia's sentient system radiate out and are part of the subtle energies absorbed by and detected along earth energy lines. There are so many factors which must be precisely so for us to exist at all, let alone flourish given our complex interconnected systems; could it really all be random?

If the Earth, as Gaia, is a sentient being, then other planetary bodies could also be ascribed as Gods or Goddesses. Mother Earth cannot be taken for granted and needs nurturing by her human children.

Today, Gaia is also the name of an ambitious mission to chart a three-dimensional map of the Milky Way, in the process revealing the composition, formation and evolution of our galaxy. This modern Gaia mission, if it can find other planets that support life, may throw some light on the key question, does our planet (via Gaia) shape life on Earth, or does existing life on Earth shape our planet?

What do the Super Clues embedded in the concept of universal and earth energies mean for our reality?

---

[17] Lewis Thomas' *The Lives of a Cell* was originally published as a collection of essays in the *New England Journal of Medicine*, 1971-73.

Summary
If the Gaia hypothesis is true, we must hope that Mother Earth does not decide she is better off without humanity and cleanse herself of us. I suspect she has done this several times before. The reality is that life on Earth can only function because of a complex self-regulatory system. The Sun's energies drive our ecosystem and enable the regenerating life cycles of nature. The Sun, Moon and cosmic forces work in harmony to make the Earth viable for humanity and all animal and plant life to thrive.

We are protected, via the heliosphere, from some of the debris floating around in space. Gravity is just strong enough to hold us in place on Earth without compromising our existence. The magnetic field and earth grids protect us from charged particles that stream out from the solar wind. Our oceans are pulled by our Moon twice daily and moderate the temperatures over the globe, allowing life to flourish more easily. So many factors need to be in position and work together, I feel it cannot be a coincidence.

There are anomalies in the magnetic grid that create vortexes or anti-gravity spots, which just could be linked to otherworldly dimensions. We are all linked energetically as well as physically, as evidenced by quantum mechanics. There are hints that the natural energy grid of the Earth has been used as a power source by ancient races or alien visitors in times past. We are not simply made up of a group of biological cells but are interactive beings, part of a vast cosmic energy system.

These Super Clues point to our planet being connected to other worlds and dimensions. Can all this really be the result of evolution, or is it part of a deliberate design? The Earth itself and other heavenly bodies may truly be sentient beings, maintaining the ecosphere to enable life to survive and thrive. If so, this gives us a whole new window on the totality of the magnitude and meaning of reality.

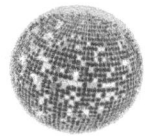

CHAPTER 7

Super Clues embedded in Alien Beings and Other Dimensions

ARE WE THE ONLY BEINGS in our reality? This is a fundamental question that has been asked very many times over the millennia. Well, from an official government and scientific perspective, humans are supposedly the only beings who exist on this planet. There is, however, an enormous amount of anecdotal evidence to indicate that other kinds of extra-terrestrial beings are here with us.

Many people, throughout history, have recorded seeing, or even working with, non-human entities. Those beings that appear to come from other worlds in our universe are known as aliens. Not only do they seem to be interacting with us in our reality, there are clues that ancient aliens have interfered in and helped shape our civilisation from the very beginning. Reports of strange creatures appearing from the sky and moving among us have, at various times, been called gods, giants or monsters.

Before we continue, I need to ask the obvious questions. Is it possible that beings from beyond the stars are visiting our planet? Who or what are aliens and what are they doing here? Do they really exist? Assuming they are interacting with and influencing life on our planet, what does their very existence reveal about the true nature of our reality?

Alien Communication
Many people believe that aliens are here and may be both helping and interfering in human affairs. Strange craft and unusual lights have been reported in numerous locations worldwide. Many questions need to be asked. Specifically, if alien beings are among us, why do only some people see them? Why, after decades of sightings all over the world, are they not in plain sight? If they are here, then why are they not officially

recognised and communicated with? These questions reach into the very heart of our understanding of reality itself.

Scientists exploring the possibilities of alien life often refer to the Fermi Paradox, named after the Italian physicist Enrico Fermi. This is the apparent contradiction between the lack of evidence that alien life exists and the high probability, given the size and intrinsic nature of the universe, that it does. It is likely that some form of life, not necessarily as we would recognise it, is an inherent parameter of the universe.

Given that alien life *should* exist, then where are they? Why can't we see them? There are many possible reasons. It may be that our human brains do not recognise other life-forms as valid images, and so do not register them. This cannot entirely be the case though as there are many tales of unsuspecting people having experiences with extra-terrestrials quite out of the blue and seeing them, sometimes very clearly. Too many sensible people, including friends of mine, have seen 'other' beings. There are, of course, always cranks and fantasists, but there is still a large body of documented incidents that cannot be explained.

It must also be asked why the official position is one of denial and files on this kind of phenomena are often classified or withheld? Many believe we are deliberately being held in ignorance. Any revelations would 'rock the boat' and may cause fear and panic. Our society is brainwashed to accept the view that we are alone in the universe, Earth is a stand-alone planet, even though we may, in fact, be part of a wider universe. A more sinister explanation is that many countries are using alien forms of technology to develop weapons or commercial materials. An even more ominous reason is that we are being ruled covertly by an alien race or races. Those who run our planet may not be who we think they are. So, are they just not here, or could it be an establishment cover-up? Maybe they do not want us to know they are here.

My sense is there are nine (at least) possible reasons for alien failure to communicate:

1. Extra-terrestrial life does not exist. They are simply not here at all and are figments of human imagination.
2. They are trying to communicate but constantly fail to do so owing to a variety of problems or mismatch of technologies; either the technology they use is far too advanced to be detected by us, or our technology is insufficiently advanced to be noticed by them. It could be a matter of frequency.

3. We are looking for extra-terrestrials in the wrong places. The universe is so vast and our equipment so relatively limited, we may just have been unlucky so far.
4. Alien life forms are so different to humans that even if in plain sight, we fail to recognise them as sentient beings.
5. Our limited sense organs may just not be detecting them, although some people do believe they are in telepathic communication.
6. They really are here, but those who run our planet do not want us to know about them for their own reasons.
7. They are avoiding us because we are bad news; we are either too aggressive or we are destroying our planet.
8. They are here but have very good reasons not to make themselves more widely known. Why this should be depends on the nature of these beings, why they are here and the purpose of the galactic federation, of which we may be a part.
9. A superior race may have a policy of non-intervention in other less advanced worlds.

When looking at whether we have been visited by alien beings from other planets, perhaps from beyond our galaxy, the general public will tend to ask where is the hard evidence? Believers would say, it is there if you look for it.

Many people not only believe they have seen or communicated with other beings but have been abducted by aliens, especially the so-called 'greys'. There is a vast body of literature by those who have researched this subject and taken eye-witness statements.[18]

UFO Phenomena
Reports of sightings of unidentified flying objects (UFOs), accounts of communication with aliens or extra-terrestrials (ETs), and reports of abduction including time lapses and human experimentation, are all more common than you might think.

Here is just a snapshot of sightings of UFOs from the vast amount of evidence available:

---

[18] See, for example, Susan Blackmore's 'Abduction by Aliens or Sleep Paralysis?' *Skeptical Inquirer*, Vol. 22, No. 3, May/June 1998. https://skepticalinquirer.org/1998/05/abduction_by_aliens_or_sleep_paralysis/

- In 1944 non-aggressive 'foo fighters' were reported buzzing and flying in formation with allied planes during the Second World War. Whether these were balls of light UFOs or prototype enemy weapons, no one seems to know to this day. It was rumoured at the time that the Germans were attempting to produce a type of magnetic weapon.
- In 1947 the first sightings of flying discs began to be reported in America. There were over 1,500 reports of daylight sightings that year.
- In the famous 1947 Roswell case, reports of a spaceship crash and dead aliens was officially maintained to have been a crashed weather balloon.
- Air Force officers who reportedly saw strange lights were told they had mistaken fireflies trapped on their windscreens.
- In 1950 a famer in Oregon took what are among the best-known UFO photos, which continue to be debated to this day.
- Sightings in 1966 of a UFO landing in Michigan were dismissed as swamp gas.
- A close friend told me of a pilot acquaintance in the South African Air Force who often saw strange objects in the sky that simply should not have been there.
- In the 1970s there were so many alleged UFO sightings in an area of Dyfed, Wales, it became known as the Broad Haven triangle.

These are just a few examples – there are many, many more. Credible witnesses are time and time again ignored, hushed up or discredited. As with any non-mainstream phenomena, people are, in the end, sometimes too scared to tell others or afraid to report what they have seen, for fear of doubt, standing out from the crowd, or concern about upsetting the status quo. Others, like the Air Force officer and his colleagues, learnt to say nothing for fear of ridicule, demotion or even losing their jobs.

The sheer volume of recorded sightings of disc-shaped objects in the sky, strange lights hovering and beaming down, and other celestial phenomena by sensible members of the public, especially those who had no interest until they saw something, surely cannot be ignored. What other possible explanations could there be for what is going on? In some instances, it may just be a covering up of government's defence experimental craft, or it may be something completely different. We cannot really know.

Why are these things not reported in the media if they are still happening? For instance, UFO sightings were extensively covered in the American media at one time, and in England articles on flying saucers appeared in magazines. Here, there are similar stories, especially in Wiltshire, Berkshire and other counties known for their military bases. The topic seems to have been dropped, ridiculed or discarded as too silly. It is, however, reported extensively on the Internet and in specialist publications. Yet the mainstream press seems to ignore such phenomena these days. There have been very few airings of an alien show on mainstream TV other than science fiction, and the American 'Ancient Aliens' documentary series.

One must remember that the news reported to us is never the whole story. Topics are picked up and dropped by the press according to news value and current trends. Considered cranky or dangerous by the powers that be, the UFO topic is, with occasional exceptions, ignored. It is worth bearing in mind that the media empires are controlled by a very few magnates who determine what subject matter and content is aired, or more to the point, what is not. The reason may be more sinister, though. The subject may have been officially suppressed.

With all the anecdotal reports, many of which are considered circumstantial and unreliable, but with little mainstream evidence coming forward, finding out the truth is problematic for those who maintain aliens are indeed among us. Throughout history people have always seen strange craft, and in more modern times reports seem to be either dismissed, distorted or covered up. The truth may, nevertheless, be stranger than fiction.

Talking to people about possible alien or UFO encounters, a surprising number of people claim to have seen something but are almost in denial about what they may have seen. They use such phrases as, "I might have imagined it," or "I was probably dreaming." I heard one woman say, " I have always seen things but am afraid of being called bonkers." This is particularly true for older people who have been chastised as children for allegedly telling tall stories or making things up. There are also those who have seen something strange but feel too embarrassed to mention it, yet they are almost relieved when they can get it off their chest to someone who does not automatically judge or doubt them.

For every person who reports a sighting or an encounter, there are others who deny the very existence of such phenomena. They claim that witnesses are misreading natural events, are fantasists or have too vivid an imagination. Then there are those who will never believe in such happenings until a little green man is interviewed on mainstream 'News at 10'! The difficulty in making sense of this subject is that hard facts are impossible to establish. There are many witness statements in books going back decades, such as John and Anne Spencer's *Fifty Years of UFOs* (1997), and now on the Internet, but there are no filters. Who is making it up? Who is telling the truth? Who is deliberately spreading misinformation? All we can do is look at an accumulation of evidence and take as much personal testimony from credible sources as possible.

As Meredith Sabini, editor of *The Earth has a Soul* (2002), reminds us, psychoanalyst Carl Jung, writing over 50 years ago, pointed out an unseen link between consciousness and matter. In other words, we may be manifesting some of these strange phenomena ourselves, through the 'collective unconscious' – a term used in his work *The Archetypes and the Collective Unconscious* (1959) to refer to the transcendent structures of the unconscious mind. Characterised by archetypes shared among us, these patterns are not shaped by personal experience. This opens-up another level of debate.

Types of Alien

There seem to be two main types of alien: those living in a three-dimensional world as we are, and other fourth-dimensional beings not of our universe. Some are said to be helping us and some are working against our best interests. Researchers who have studied the phenomena identify several different types, shapes and sizes of alien beings said to be among us, along with different kinds of craft. While impossible to prove or disprove, the list of alien species recorded to be in contact with humanity runs into many hundreds. They range from several categories of Greys, Nordics, Insectoids and Reptilians to, indeed, "little green men", usually depicted with large bald heads and huge staring eyes. Some may even be silicon based and have completely different DNA, which could explain why, according to abduction stories, they are so interested in us.

Is there a level of predictability to evolution that would cause aliens to look more like us than films and science fiction would have us believe? Beings that evolved in an interstellar dimension would, of necessity, have very different needs from those who come from an Earth-like planet. As for beings not of our three-dimensional universe, seeing them would be more difficult. The problem is the matter of vibrational frequencies. Limited by our sense organs and brains, we simply could not see a complete four-dimensional craft, only in three dimensions. It would partially appear and then suddenly seem to vanish, which accords with some reported sightings. This may also explain strange lights and shapes seen in the sky which then disappear. Science does tell us that we live in a multidimensional universe or even a multiverse, after all. Although visibility is difficult, telepathic contact still seems to be possible, according to those who allegedly have had direct contact. The bodies of four-dimensional beings are vibrating at a higher frequency than ours. Our sense organs are not tuned to detect them, although they can see us at a slower frequency. This, of course, adds further possible reasons for our failure to communicate. They may be all around us and not look anything like humanoid, but we do not recognise them simply because they remain invisible to us.

My own researches and people to whom I have spoken all suggest that many aliens are concerned about the welfare of our planet. They are worried about what destruction we may be doing to the balance of nature. It is also rumoured that many of our nuclear accidents and global disasters might have resulted in more severe destruction, or been avoided altogether, had they not in some way been tempered down or reactors switched off. Are we being monitored and helped at critical times, without being overtly interfered with?

Another area where there seems to be a more sinister component is alien abductions. Many people feel they have been abducted and medically interfered with, even 'transported' from their own beds. There are also those whose cars suddenly won't start and/or find themselves with unexplainable time-lapses.

Probably the best-known case of alien abduction is that of Betty and Barny Hill who, in September 1961, were stopped on the road in New Hampshire by strange lights which seemed to follow them. When they finally got home dishevelled, their watches no longer working, two hours of the journey had elapsed that neither of them could remember. They had been 'kidnapped' by aliens and taken inside a large metal disc, where the beings had examined them and erased their memories. Regressive hypnosis played a major part in enabling the couple to recall what actually happened.

There is even talk of hybrid alien-human experimentation. There have also been strange cases of farmers reporting mysterious cattle mutilations at the time of UFO activity. These bloodless incidents, affecting a great many animals, have allegedly taken place in America, Argentina and Australia. Other than speculation that it was done for experimental purposes, no convincing explanations have ever been given.

There may be alien races from other planets or dimensions in the universe, both helping us and interfering in our reality.

There is, as mentioned, a deep suspicion of a cover-up and/or collusion going on. Are our governments covering up alien interaction, for reasons that may or may not be in our best interests? Well, some people think they are, and co-operating with aliens in exchange for their technology. Others believe that even if they were, humans would have great difficulty in understanding, let alone reverse engineering, such advanced technology for weapons or other purposes.

Chris Harris sent me his thoughts, explaining the issue succinctly,

"All major governments are of course in touch with alien groups that are visiting our planet. It is the science and truth known to these travellers that prevents disclosure. To have a full understanding of the mechanics of space-time and to understand the reality behind each of our souls' interconnectivity and immortality is to know more than can be released to Earth humans at this time. To unleash revelations exposed by the exotic science of our alien contacts, is a logistical nightmare that governments are trying to address. It is not panic of the masses at stake, it is the sanity of us all. Things now so important to us that cement the world together would dissolve if our education is too fast."

UFO Sightings
Let us summarise well-known reports relating to UFOs and aliens, in both America and in England.

- **The Roswell incident**
  The Roswell incident is another contentious subject. While 70 years of conspiracy theories may have clouded the story, what is agreed

is that an unidentified flying object crashed on a ranch northwest of Roswell, New Mexico, sometime during the first week of July 1947. What it was is a matter of speculation. A rancher is said to have witnessed this happening and picked up debris over a large area, made from materials he did not recognise. The military reportedly "shooed away" anyone attempting to look at the site. The authorities later said that it was just a weather balloon. It may also, of course, have been a human 'top secret' project.

Donald R. Schmitt in his book *Cover-Up at Roswell* (2017) and Kevin D. Randle in *A History of UFO Crashes* (1995), both claim to have uncovered a conspiracy. Their research shows that military radar had been tracking an unidentified flying object in the skies over New Mexico for four days. On the night of July 4, 1947, radar indicated the airborne object had gone down on a ranch about 30-40 miles northwest of the town of Roswell. It was subsequently recovered. What's more, former mortician, Glenn Dennis, claimed in 1989 that a friend who worked as a nurse at the Roswell Army airfield had accidentally walked into an examination room where doctors were bent over the bodies of three creatures. They apparently resembled humans, but with small bodies, spindly arms and giant bald heads. Truth or fiction? So many people believe what crashed was an alien spacecraft with three bodies inside, this seemed to spark the belief, rightly or wrongly, that the government was covering up the facts about our interaction with aliens. There is no way of knowing. In its lengthy report, released in 1997 (when the 50-year statute of limitations had expired), the US Air Force denied all of it.

- **Area 51**
Roswell is not far from a facility known as Area 51. Code named Dreamland, this is the radio call sign of a large 'top secret' military Air Force base.[19] Located in the southern portion of Nevada in the western United States, 83 miles (134 km) north-northwest of Las Vegas, what goes on there the US government will not discuss. It is so secretive that employees are made to sign a lifetime secrecy document. The size of the base is six miles wide by ten miles long.

---

[19] See Dreamland Resort, *Secrets of Area 51 Revealed*, 'FAQ: What other names are used for Area 51?' http://www.dreamlandresort.com/faq/faq_other_names.html

Apparently, the federal government has seized an additional 85,000 acres surrounding the base to keep observers at a distance. It is said that civilian contractors and other military personnel at the base are flown in every morning aboard unmarked planes with blacked-out windows. Structures are allegedly built deep into the desert floor, with buildings going down 30-40 storeys below the surface. It is alleged that specialised work is done here to backwards engineer alien technology from crashed spacecraft. A more sinister school of thought suggests it is in fact a centre for alien-human co-operation, where experiments into cloning hybrid alien-humans were or are being done. Tellingly, there is a 'no fly' zone above the area, extending as far as the edge of the atmosphere.

There is a strong suspicion that the US government were using the cloak of UFO sightings to cover up the testing of its own experimental aircraft. This was especially true in the Cold War. At the same time, it was ridiculing the phenomena. Why? One explanation is, quite simply, they could not admit to using alien technology as in doing so they would lose their defence advantage. Were there also 'real' alien craft visiting? Was this just a gigantic double bluff? It is impossible to tell. Also, if national governments were to admit to the presence of alien visitors, it would change the status quo and undermine self-interest. There is also supposed to be another secret base, 33 times bigger, called Area 32, in Utah. It is suspected that much of the most secret work has been moved there. The truth is very hard to discern, but there is definitely something very strange and highly secretive going on.

- **Rendlesham Forest**
  Between two UK military RAF bases near Ipswich, Suffolk, officers on security patrol were investigating what they thought was a downed aircraft, when they saw unusual beams of white light illuminate the forest. Two nights later, on Christmas night 1980, strange strobe-like coloured lights flash before them and disappear, but not before a triangular silver object with three legs, a pulsing red light on top and a bank of blue lights underneath, had landed. The bases at Bentwaters and Woodbridge were both being leased to the US Air Force, with credible witnesses to this mysterious incident.[20]

---

[20] See http://www.therendleshamforestincident.com/

When Josh Gates, American host of *Expedition Unknown* aired on the Travel Channel since 2015, years later interviewed the commander about what he saw, I watched him say that high radiation readings were taken in the area. He confirmed all tapes of radar evidence were removed by the MOD and serving officers were told never to talk about it, as if it had never happened. In 2002, the UK MOD released the document file on the Rendlesham Forest incident, dubbed by the media as 'England's Roswell'. Nothing was reported beyond lights from a windmill!

Testimonials

One way of making some sense of it all is to talk to trusted friends who have had a personal experience. This first account of an unexpected encounter with a UFO that took place nearly twenty years ago, was told to me very recently. It is significant not just because it was not looked for but, importantly, because it changed the person's outlook on life.

The second account is testimony by Whitley Strieber, describing his interactions with and abductions by 'grey' aliens. Although over thirty years old, it still has important points to make.

- **UFO testimony – Genna**

    Genna and her friend stopped briefly on their way home from a party, tired but sober at 11.30 p.m. at night. They pulled into a car park in the village of Sutton Scotney, Hampshire, which overlooked a recreation ground, to visit the toilet. Suddenly, they saw a cigar-shaped pulsating ball of fire above the park. Mesmerized, they sat on the bonnet of the car watching. It then moved closer and looked like it tripled in size. It seemed to be continually changing position and form. Occasionally, a saucer-shaped craft could be seen beneath the outer pulsating glow, with blue and red lights. It then beamed light downwards, behind the trees in the park. Then two flashes and it was gone. This experience lasted twenty minutes and felt peaceful, but in Genna's words, "it changed my life".

    Her description of the craft matched what many people have seen, all over the world. Twenty years ago, much of this knowledge was not in the public domain, and Genna was certainly not privy to, or interested in, the subject of UFOs. Some people may feel that what she described is similar to anti-gravity technology, with some sort of cloaking device around the spaceship.

What I found especially interesting is the fact that Genna felt this experience changed her life. She came from a strict religious background, and this encounter threw her life into disarray and made her doubt her faith. She felt great anger at people's level of disbelief when she shared her story. She felt almost an outcast, separated from her friends, who did not believe her. It made her question things on all levels. Had she seen demons? What else had been hidden? What were the aliens doing here? These thoughts were clearly on her mind. For someone of only 19 years old, the aftermath rather than the actual experience was the ordeal. She did not expect to see such a thing and it changed her life, for the worse to begin with. Yet, it was purely an experience of seeing an alien spaceship.

- **Abduction testimony – Whitley Strieber**
  What I find most interesting is Strieber's doubts, even as the bizarre events during his physical abductions with aliens around his bed, are happening. He thinks he may be going mad, or that these experiences may be coming from his subconscious mind. His book *Communion* (1987), later made into a film, explains periods of 'lost time', during which he feels that he interacts with four types of 'grey' species. The small common ones he believes may have a hive mind, as they certainly sometimes move as one – some now believe these to be clones. The others, while looking vaguely human, albeit slender with no muscular definition, give him an insect-like impression. All have these egg-shaped heads and vast, dark eyes. He has the sense of having always interacted with one particular individual who communicates with him. He says she is very old. She reminds him of the goddess Ishtar.

    Where do these beings come from? He gives a list of possibilities, some of which I had not considered before:

    - From Earth – another form of human species so different from us that we had not until now known they existed or were real.
    - From another planet or planets.
    - From another dimension of space-time.
    - From this dimension in space but not time.
    - From within us – we are creating them with our minds.
    - A side-effect of natural phenomena, certain magnetic frequencies tripping a hallucinatory wire in the mind.

Strieber's experiences appeared to stimulate other memories that his conscious mind had previously no recollection of. The effect this had on his identity made him realise that his forgotten life was more 'real' than the one he was presently living. As an already published author, he was, however, put in the position of having to defend his story as not being entirely fiction.

These two testimonies are just a small part of the information and experiences which are out there. There are hundreds and thousands of first-hand witness reports and personal accounts to be found in books, articles and on the Internet. While some of them may be false or fabricated, they cannot all be untrue.

Intergalactic Intervention
My many discussions with people who have been investigating the concept of alien contact and intervention has thrown-up several disturbing concepts, all of which point to the possibility that we are not as free and independent as we may think we are. Matters needing further investigation include the following possibilities:

- Humanity was genetically altered by aliens for its own purposes at the beginning of time.
- Humanity is and always has been controlled, covertly, by our alien masters in disguise or by their representatives.
- There are alien races working in our best interests and trying to help humanity counter those that are not.
- Humanity may be likened to a virtual reality programme; either side may have hacked into the others' game.

There is a possibility that we may have been genetically manipulated at some time in the past and are unknowingly controlled by alien beings.

Beyond stories of aliens interacting with individual humans, there is allegedly a level of interaction on a much greater cosmos-wide scale.

- **An intergalactic federation**
  A surprising number of people believe there is an intergalactic federation of alien races, a large and powerful conglomerate devoted to universal peace and prosperity. It consists of alien members of

planetary civilisations from many different galaxies and universes, working together for the harmonious existence of all life. Every inhabited galaxy supposedly has its own federation of light. The overall structure is vaster than we can imagine. The forces of darkness have, so far, been prevented from getting the upper hand.

- **Malevolent aliens**
Some people believe there exist members of a malevolent extra-terrestrial (ET) confederation. This consists of reptilians from Alpha Draconis, aliens of the Orion group, and greys from Zeta Reticuli. It also appears that the reptilians have been manipulating humanity into servitude for aeons. They have been feeding off our labour and using our human bodies through the wars they instigate and the hostile belief systems which they foster, often in the form of controlling religious and social institutions.

- **Intergalactic portals**
This is a fascinating but mind-blowing concept. Chapter 6 touched on portals built into the structure of the Earth. Finding out what is really going on is impossible. Rumour has it that NASA and other government agencies are researching and even using interdimensional portals to teleport over the Earth, and to link into intergalactic highways. These are said to be used by alien visitors to Earth and vice versa. There is no proof of this actually happening, though it is an intriguing notion.

Some confusion may arise, however, over NASA's discovery of hidden 'portals' prior to 2012 in the Earth's magnetic field, termed x-points or electron diffusion regions, which is not the same thing.

There may be interdimensional portals built into the Earth which allow travel between alien worlds and other dimensions. Teleportation is also a possibility.

To offer another perspective, Andrew Collins, in his book *LightQuest* (2012), suggests that UFOs are the product of intelligent light forms, plasma based inter-dimensional intelligence that interacts with our consciousness on some sort of archetypal level. Such sentient energy forms and plasma constructs may be manifestations of higher dimensional realities and could even be a part of Earth's evolution.

Ancient Aliens

There are many myths and stories of alien visitations in our ancient past. These may tell us the origins of our civilisation as we know it and have been taught, is completely wrong. We may not be who we think we are. Where we come from may not be planet Earth at all. Some people believe that alien races, such as the Annunaki, are inextricably tied up in the history of our species and are still with us today.

Encounters with aliens are not a new phenomenon. They have been happening throughout recorded time. Ancient legends give us clues. Alexander the Great allegedly reported an attack on his army, in 329 BC, by two flying objects, which Frank Edwards, in *Stranger than Science* (1963), tells us were described as:

> "great shining silvery shields, spitting fire around the rims ... things that came from the skies."

Not only do the ancient Chinese mention flying carts or winged chariots, flown by one or three-eyed men, there is the much-quoted biblical vision of the prophet Ezekiel of a vessel, or chariot, that descended:

> "And every one had four faces, and every one had four wings. ... and they sparkled the colour of burnished brass."[21]

What is Ezekiel describing? Erich Von Däniken believed it to be a UFO. Josef Blumrich, NASA engineer and author of *The Spaceships of Ezekiel* (1974), set out to try and refute his interpretation, but could, surprisingly, see from the description a viable spacecraft. Even though some of Von Däniken's work is recognised as flawed, owing to mistranslations of two different Bibles in different languages, as a non-biblical scholar he made a serious study of *Ezekiel* which sparked his interest in UFOs. Ezekiel's vision became one of the most important texts in Jewish mysticism, as God's chariot on which Ezekiel rode to Heaven.

Are these stories just myths, interpreted with modern eyes? Or are they a good indication that we have been visited since time immemorial by alien races? I revisited the works of Erich Von Däniken, whose

---

[21] *Ezekiel 1:5-10*. Holy Bible, King James Version.

arguments presented over 40 years ago, in his books *Chariots of the Gods?* (1969) and *According to the Evidence* (1977), are still relevant today. It is his contention that, in earlier times, the Earth was visited by unknown beings from beyond our galaxy. These visitations were recorded and handed down through religion, myth and legends. Many ancient stone carvings, in cultures all over the world, also show objects and figures that could be interpreted as spaceships and aliens in space suits.

Others support the view that aliens created intelligence in our hominoid ancestors by genetic manipulation. Our Homo Sapiens' city of civilisation supposedly began at Sumer in Mesopotamia, around 6,000 years ago. This date is surprisingly late. As we saw in the previous chapter, it was James Hutton, in 1785, who proved the Church of England wrong in their date of the beginning of creation.

The alternative narrative is that about 432,000 years ago, an advanced civilisation from a mysterious tenth planet beyond Neptune, with a 3,600-year elliptical orbit in our solar system, known as Nibiru, splashed down in the Persian Gulf. This brought the race which the Sumerians called the Annunaki. According to the controversial author Zecharia Sitchin, the Annunaki discovered and began to mine the gold they needed to help repair their atmosphere. He tells us in the *Annunaki Chronicles* (2015), that in about 250,000 BC their miners apparently revolted, and their leader, Enki, decided to make early humans more like them and genetically manipulated the local population to create a slave race. When they merged primitive Homo Erectus with their own genes, the offspring became the Nephilim, referred to in *Genesis* as "the men of renown."[22]

Called 'giants' in the King James Bible, the Nephilim bred and became numerous and spread out over the rest of the planet. Most of this population was wiped out by the great flood when planet Nibiru returned close to Earth. However, enough were saved to begin again. Before returning to their own planet, the Annunaki helped to set up, teach and guide the emerging civilisation in the Middle East. Their presence is documented on stone tablets in the city of Sumer. Were descendants of these beings responsible for building our ancient structures and pyramids? The Sumerians were, incidentally, among the first people to leave sophisticated astronomical records, aware of the Pleiades in marking the beginning and end of the agricultural year.

---

[22] *Genesis 6:4*. Holy Bible, American Standard Version.

Although this scenario is considered by some to be the story of our origins, many scholars disagree. They suggest the tablets tell a completely different story and the Annunaki were no more than gods in early Middle Eastern civilisation. Some even believe they were 'Watchers', or angels, sent from God to help early man. Tall tale? Perhaps, but all I can say is the jump from man-ape to complex and abstract thinking is a difficult one in purely evolutionary terms. Nevertheless, the closest living relative to Homo Sapiens is the chimpanzee. So, if this is true, then we all have alien DNA in our genetic makeup; we are hybrid-hybrids. We are all from the stars.

It is worth mentioning that the planets were known as gods in ancient times, as evidenced in the Roman and Greek names: Mercury (Hermes), Venus (Aphrodite), Mars (Ares), Jupiter (Zeus) and Saturn (Cronos). Together with the Sun and the Moon, these were the seven classical planets of antiquity.

We may have been seeded by alien beings that could even have been the gods referred to in ancient times. Alien intervention from beyond the stars may have been vital for our development.

Stories from Around the World
There are stories of extra-terrestrial visitations and ancient alien cities from all over the world. South America has a rich history both in ancient aliens and UFO experiences, which continues to this day. Many of its remains include precision-cut granite stones, some weighing over 10 tons, laid accurately in place to tolerances almost impossible to achieve today. In Egypt and the Middle East, similar clues indicating possible visitations are left in the surviving stonework of ancient cities and temples. Africa and Australia have their own stories. These are just some examples of how we have been visited and interfered with, or helped, by otherworldly beings throughout the ages.

- **South America – the Nazca lines**
The Nazca lines in Peru are often mentioned as proof of alien contact. There are massive and extensive geoglyphs based on figurative animal, plant and geometric shapes and lines, constructed on the desert floor, southeast of Lima. They have been created by removing the top layer of soil and exposing the whitish rock of the dry plateau. The largest of these stretch almost 1,200 feet across and can only be

properly seen from above in a plane or spacecraft. Some resemble runways and are built to an accuracy that would seem impossible to achieve without detailed plans or a birds-eye view. One singular design, known as 'el astronauta', does look rather like a man in a spacesuit. These lines were once said to have been made by the primitive Nazca people, who are believed to have disappeared around 500 AD. It is now suggested they are very much older.

I watched two recorded television programmes about these mysterious lines in quick succession. In one documentary, a group of scholars deduced they were extensions of a Nazca temple complex and used in rituals. Another stated that their very nature almost certainly meant there had to be an otherworldly hand (alien) in the construction of at least some of them. It was also pointed out that the extensive lines point east to the megalithic cities, which were abandoned around the time the Nazca vanished. Skulls found at these sites were large and elongated, as found elsewhere.

- **South America – Puma Punku**
  No discussion of ancient aliens can ignore the ruins of the ancient city of Puma Punku, meaning "door of the puma", located 13,000 feet up in the Bolivian mountains. Part of the larger archaeological complex of Tiahuanacu (Tiwanaku), and initially thought to be about 2,000 years old, Puma Punku is now believed to date from at least 12,000 years earlier. An article published in 2019 by *Ancient Origins*,[23] explains that archaeologists found the astronomical alignments of the main temple to indicate that it was built to coincide with the summer and winter solstices and spring equinox, as would have been seen 17,000 years ago.

  Pre-Inca myths describe how the Earth and all living beings were formed by the great creator sky god, Viracocha. In his book *The Lost Tomb of Viracocha* (2001), Maurice Cotterell retells the legendary story that he appeared with his giant servants through the famous megalithic stone 'Gateway of the Sun', at Tiwanaku. Viracocha is said to have first created a race of giants from stone, but when they became unruly he sent a great flood to destroy their city. Next, he created men and women from clay, and gave them the gifts of

---

[23] Alicia McDermott, 'Puma Punku: This Ancient Andean Site Keeps Everyone Guessing', *Ancient Origins*, June 2019.

language and agriculture. Then he made the sun, moon and stars to bring light to a dark, stone world, before setting off around the Pacific to spread the knowledge of civilisation. When he left, lesser deities were assigned the duty of looking after mankind. Could these gods have been ancient aliens?

- **Africa – Sirius and the Dogon**
  The Dogon are a remote tribe living in a desert region of Mali, thought to be of Egyptian descent. Their legends date from before 3,200 BC. Laird Scranton, in *Science of the Dogon* (2006), explains that in 1931 they were contacted by two French anthropologists, who studied them for thirty years and documented the tribe's traditional mythology and beliefs. This included a body of ancient star lore regarding Sirius (the Dog Star). The Dogon were aware that:

- Sirius had a companion star, which was invisible to the human eye, Po Tolo.
- This star moved in a 50-year orbit around Sirius.
- It was small, incredibly heavy, and rotated on its own axis.
- There is a third star in the system (the second companion), Emma Ya, which has yet to be identified by astronomers.

The facts are that Sirius B does exist. It is a very heavy white dwarf that rotates on its own axis. It does have a 50-ish year elliptical orbit around Sirius B. So how could a primitive tribe know all this? Robert Temple, in his book *The Sirius Mystery* (1998, first published in 1976), provides convincing evidence that the Egyptian, Sumerian, and Dogon civilisations were founded by aliens from the Sirius star system. The author tells us about the nature of this star, only eight and a half light years away, which was not seen though a telescope until 1862 and only photographed in 1970. Artefacts found describing the Sirius star system are at least 400 years old. The Dogon say this information was given to them by the Nommos, amphibious beings sent to Earth from Sirius for the benefit of mankind.

Other ancient cultures mention these fish-like gods. The Babylonians called them Annedoti, who gave them the basic teachings of how to organise a civilisation. This parallels the pre-Inca myth of Viracocha. There are many other similar stories, suggesting that alien intervention may have been with us right from the beginning.

- **Ancient Egypt**
  Although there are many stories of their pharaohs, the Egyptians were largely an agrarian society with a powerful and organised ruling elite. It is not easy to establish what happened in ancient Egypt 5,000 years ago, when their monumental pyramids, temples and obelisks were built. How did they manage these enormous feats of engineering with no modern construction tools? There is a growing belief among historians, despite the general view that the pyramids were built by primitive construction methods using intensive peasant labour, pulleys and stone rollers, that advanced technology may have been used.

  The three main pyramids of the Giza plateau are said to be aligned to the three main stars of Orion's belt, a constellation that had religious significance for the Egyptians. While others disagree, this theory was originally proposed by Robert Bauval and Adrian Gilbert in *The Orion Mystery* (1994). The pyramids are believed to have been sited at the exact latitudinal and longitudinal centre of all land areas on the planet, with such mathematical accuracy that the height of the Great Pyramid correlates with the distance between the Earth and its sun.

  The Great Sphinx is another mystery. Oriented on a straight east-west axis, it was known to the Egyptians, who venerated Horus in his role as a sky god as Horemakhet, 'Horus of the Horizon'. Its purpose continues to generate much debate among scholars. It is possible that the Sphinx already existed when Khufu came to the throne in 2589 BC and began to build his Great Pyramid in this once fertile region with abundant underground aquifers.

  Some archaeologists now believe that the pyramids, with their long, sloping chambers, were never intended to be built as tombs and are, in fact, gigantic power generating plants. The discovery of what are called the Bagdad batteries, cells thought to be 2,000-5,000 years old (depending on source, including *We are Not the First: Riddles of Ancient Science* (1972) by Andrew Tomas), capable of generating a low voltage electricity, shows that a knowledge of and use for electricity did exist in those days. When I visited Luxor Museum many years ago, and looked at some of the statues and artefacts, it was difficult to imagine them being made by anything other than power tools.

  Did they have help? Could the ancient Egyptian gods have been aliens? Or is there a simpler explanation and their knowledge died with the end of their civilisation and absorption into the Roman Empire? Clues remain carved into the temple walls. For instance, spaceships, aeroplanes and a helicopter can be seen on the walls of the temple of Seti I in Abydos. There are also gods looking very much like the bird gods found in South America and elsewhere. Depictions of what can only be described as alien-like humanoid beings were more recently found in hieroglyphic panels on the walls of other temples. It is also claimed that images likened to induction coils and light bulbs can be detected on the walls of the temple of Dendera. There will always be debunkers, yet this evidence appears to indicate that the Egyptians had superior intelligent support.

- **Ancient Vedic Myths**
Indian Vedic and Sanskrit literature abound with descriptions of flying machines. As we have seen, mythical gods are said to have travelled and ruled, allegedly waging warring battles in the skies. The Sanskrit text *Drona Parva* describes what appear to be aerial dog fights between gods in flying machines, called Vimanas.[24] Some Vimanas that resembled aircraft with wings and flew in a mysterious manner, are thought not to have been made by human beings. There are descriptions in the epic poem *The Mahabharata* of what would appear to be interdimensional spacecraft, chariots capable of very fast speed and described as not of this world.

    We all grew up with fairy tales of the flying carpets of Arabia. Could they have existed after all?

- **Australian Dreamtime**
I always felt there was some kind of mystery surrounding how the indigenous aboriginals arrived in Australia, given that history tells us mankind evolved in Africa. There was a fudging about lower sea levels but no proper explanation. I also remembered the suspicion that their culture was actually much older than ours, 60,000 years at least, by their reckoning. The problem is, of course, with no written language, it all becomes tricky. Their tradition, apart from a few rock paintings, is passed down orally in the form of stories, song and ritual dance. There has been a general reluctance in Western culture

---

[24] *Drona Parva* is the seventh book of the Indian Mahabharata.

to believe such unverifiable sources, not to mention an undeserved presumption about the primitive nature of their lives.

The traditional Dreamtime legend says the rainbow serpent held all of creation in its belly, including the first man. As an immortal being and the giver of life, it can also be a destructive force if enraged. Aboriginal legend tells that all life was seeded by sky gods, in the remote past. Could the mythical rainbow serpent have been a spaceship? In an episode of *Ancient Aliens*, 'The Wisdom Keepers', aired on the History channel in September 2017, the comment was made that when asked where they come from, they will point up to the sky and say they come from beyond the Milky Way. They also have an ancient knowledge of star lore and their relative positions in the heavens. Aboriginal cave paintings show beings wearing helmets, looking suspiciously like the famous, or infamous, grey aliens. Barry Watts verifies, in his book *UFOs Down Under* (2017), that Australia is also a well-known UFO hotspot.

Stories abound in myth, rock carvings and temples, and in ancient scriptures, that indicate the origins of humanity may be very different from what we have been taught.

Lost Technology

In examining some of the myths and legends surrounding our origins, there seems to be a common theme. Specifically, the construction of ancient cities involving enormous blocks of granite and built to fine tolerances. Legend tells us these huge blocks were simply floated into place, as conjectured in ancient Egypt and even at Stonehenge. Could this have been done using ancient technology that no longer exists?

There is no reasonable explanation of how such feats were achieved in human terms, considering the primitive cultures and limited tools (before the wheel had been invented). In some regions few trees existed to assist with construction and any roller-delivery system. Machine tools or other advanced technologies would surely have had to be involved. So, these megalithic constructions, found worldwide, may have involved either ancient technologies that have since been lost, or alien knowledge, perhaps even assistance from another dimension.

I understand there are two possible technologies known to us which could accomplish this. Firstly, anti-gravity, which involves balancing the force of gravity with some other force, such as electro-magnetism.

Secondly, by levitation, using acoustic sound waves to achieve the same thing. Willem Witteveen's *The Great Pyramid of Giza* (2016) is a fascinating exploration of the properties of sound waves, Schumann resonances and the special abilities of quartz. Research into both these practices, to try and learn more about these technologies, is still, today, in its infancy. They may, however, be complementary and not separate technologies at all, since, as we saw in Chapter 6, diamagnetism could be at the root of anti-gravity, the force which enables levitation.

There is also a theory that there may have been a way to soften stone to allow the huge blocks to be moulded into place. Some ancient South American structures certainly look as if they could only have been built in this way. Researchers Jan Peter de Jong and Jesus Gamarra refer to three distinct styles used in different cultural periods in Cusco, the old Inca capital of Peru.[25] The earliest of these had the ability to heat stone to very high temperatures, which created a vitrified and smooth surface. In this way, polygonal blocks could be placed into position and allowed to cool, creating an interlocking structure using a process beyond our understanding today. Could these technologies have come from aliens or even our own long-lost civilisations? Either possibility brings the potential for a complete re-writing of human history.

There is evidence in our ancient cultural myths of lost technologies that may have come from alien sources or a far older civilisation on Earth.

Can we ever be sure that we have any idea whether intergalactic beings existed on Earth in the remote past, and if they continue to interact in a realm beyond human consciousness today?

Before we move on, it is of relevance to mention the orbit of our solar system around the Milky Way galaxy, in cycles of what are known as Great Ages, which take 25,626 (approximately 26,000) years. Owing to precession of the equinoxes, the celestial poles shift as the Earth rotates on its axis. As the constellations appear to slowly rotate around the Earth, we enter each of the zodiac signs in turn for a period that lasts approximately 2,160 years. We are presently moving from the Age of Pisces to the Age of Aquarius. Hinduism refers to the Kali Yuga, the end of four ages of a larger cycle, which is often referred to as the "dark age" when the world soul is black, before the beginning of a new cycle.

---

[25] See 'The Cosmogony of the Three Worlds'. www.janpeterdejong.weebly.com

We are presently in the age when many people have lost their spiritual awareness, mental clarity and can turn to wickedness. We can only hope and pray that this shift will bring a new humanitarian approach.

Men Ahead of Their Time

Nikola Tesla and Victor Shauberger were both were silenced for their pioneering inventions, including free energy, which we are only today beginning to replicate.

- **Nikola Tesla**
  Born in 1856 to Serbian parents, Nikola Tesla was a powerhouse of imagination and design. It was his invention of alternating current, rather than Edison's of direct current, that powers our world today. Tesla's inventions went far beyond electricity; he envisioned concepts that were way ahead of his time. His many inventions included wireless radio communications, turbine engines, helicopters, fluorescent and neon lighting, torpedoes and the X-ray, also antigravity technology. Even more remarkably, using his Tesla or Wardenclyffe towers, he designed a method to utilise and wirelessly transmit the Earth's own geothermal energy. However, his funding was withdrawn when his backers learnt what he was doing, because they thought they could not make money from it.
  According to Marc J. Seifer, in *Wizard: The Life and Times of Nikola Tesla* (1996), in his own words, Tesla was using,

"... the Earth itself as the medium for conducting the currents, thus dispensing with wires and all other artificial conductors ... a machine which, to explain its operation in plain language, resembled a pump in its action, drawing electricity from the Earth and driving it back into the same at an enormous rate, thus creating ripples or disturbances which, spreading through the Earth as through a wire, could be detected at great distances by carefully attuned receiving circuits. In this manner I was able to transmit to a distance, not only feeble effects for the purposes of signalling, but considerable amounts of energy, and later discoveries I made convinced me that I shall ultimately succeed in conveying power without wires, for industrial purposes, with high economy, and to any distance, however great."

Had he been allowed to develop his discoveries Tesla could have powered the world for free. Scientists are only today managing to achieve limited wireless transmission of power under laboratory conditions. Clean energy companies, particularly in America, are now developing free energy devices. Side-lined and ridiculed in his lifetime, when he died in 1943, Tesla held nearly 700 worldwide patents. His body and remaining papers mysteriously disappeared. Way ahead of his time, where did Tesla get his ideas? It seems he received detailed images in his mind. Where did they come from? He was attempting to communicate beyond Earth. Could he have been a conduit for alien information, perhaps a two-way stream? Many people believe so. Was Tesla another recipient of alien assistance in our civilisation, albeit one that was essentially sabotaged by the human industrial complex of his time?

- **Viktor Shauberger**
  An Austrian, Viktor Shauberger, born in 1885, had a fundamental understanding of the life-force inherent in nature. Observing the natural flow and energetic properties of water, he developed methods for soil improvement to enhance growth and the quality of crops, paying particular attention to restoring the natural flow of rivers. Olaf Anderson, in *Living Water: Viktor Schauberger and the Secrets of Natural Energy* (1996), writes that among his many inventions was the trout turbine, based on the functioning gills of a trout. Known for his work on vortex mechanics, his observations of rivers and streams were equally applicable to air. His inventions included a model of flying saucers later built for the US Air Force. In 1934, he met Adolf Hitler to discuss the underlying principles of agriculture, forestry and water, with the view to developing projects to harness the vortex-like forces in nature. This allegedly included replicating alien technology.

  Using a technique known as implosion, like Tesla he became a pioneer of free energy. Schauberger noticed that the driving force of movement was centrifugal, or spiralling outwards, which generated friction and heat and tended towards destruction. However, if the spin was focused inwardly, the centripetal force would encourage nourishment and growth. It would also favour levitation, just as when water vapour cools and rises above as clouds. His revolutionary theories using implosion technology to find a balance between these forces could potentially lead to cleaner, sustainable energy today. Viktor Shauberger died in 1958, only five days after returning to Vienna from a trip to America, where he was made to sign an agreement that prevented him from doing any further research.

What do the Super Clues embedded in the concept of alien beings and other dimensions say about our reality?

Summary
It is unlikely that we are alone in this reality of ours. Aliens have been observed and recorded as interacting with human beings throughout history. They may have always been with us and could be here now, directing and helping us, or possibly using us in unknown ways. They may even be covertly running our planet for their own purposes.

Myths and legends abound in many ancient cultures of gods and beings from other dimensions who have both interfered with and protected mankind. There are many theories concerning the nature of this contact, both past and present. In some circles, it is thought we are being watched and monitored by an intergalactic federation, and that there is a cosmos-wide battle between good and evil taking place. We may even be living in an alien computer generated reality or a universe created by artificial intelligence.

What this says about the nature of our reality, if all or any of it is true, is that we are not the autonomous creatures we believe ourselves to be. We may be, or have been, directed and manipulated, as unknowing pawns in another's game. Any available testimony is always open to interpretation. This is particularly true when a culture has no written record of its history, yet oral tradition is always compelling. There is also evidence worldwide of ancient technologies now lost to us, and of those individuals in more recent times whose work has been deliberately curtailed.

Our perception of reality has been influenced in untold ways – all we have are clues. There may be rather more than a tantalising glimpse in the information presented here that our present civilisation is not the first on this planet. The truth about our ancient past, and what is really going on today, is very difficult to pin down with any certainty. However, the sheer bulk of anecdotal evidence makes it hard to dismiss completely the idea of alien beings from other dimensions and galaxies, playing a role in our reality.

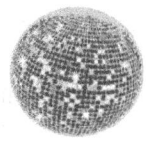

CHAPTER 8

Super Clues embedded in the Spiritual & Angelic Realms and Orbs

ANGELS ARE SPECIAL. They are entwined with our sacred and secular traditions, and are found in both Eastern and Western religions, mainstream and alternative cultures and practices. Of all the other beings in our reality, angels are perhaps the most familiar and are certainly the most widely illustrated. Who or what are they? What do they look like? What are angels really doing here with us?

The truth of the matter, as ever, depends on what the big picture of our universe actually is. There are differing opinions as to what angels indeed are, which seems to depend on individual beliefs about the nature of our reality. There is the purely religious view of the existence of angels and their place in the hierarchy, and alternative ideas that relate to the nature of our existence and our place in the cosmos.

Angels
People across all religions and with all kinds of personal experience perceive angels differently. This is, of course, because, to some extent, we see what we expect to see. Our brains, on seeing something new, will search our mental database and show us what we imagine an angel to look like. When Renaissance painters began to portray angels with wings and human-like bodies, this is how many people saw them. Some people see the expected white-winged angels in human form, while others see them as energy beings.

Many people claim to have personally seen or to have had contact with angels. Modern literature and the Holy scriptures in Christian, Jewish and Muslim religions, abound with stories of angelic intervention on behalf of humanity. The best-known Christian story is by the apostle Luke, who recounts when the Angel Gabriel appeared to the Virgin Mary, to proclaim she would give birth to Christ the Saviour.[26]

Story after story tells of personal angelic intervention in the lives of ordinary people, usually those of strong faith, but not always. George Washington was shown a vision, in 1777, by an angel of the union of the United States of America, later known as the 'angel of liberty'.[27] Many artists and poets claim to have worked with and been inspired by angels. William Blake, for example, had visions of angels from the age of eight, in the 1760s, which are illustrated in his *Songs of Innocence and Experience*.

Angels are also seen at bedsides when people are about to pass over. An acquaintance recently told me of a strange tall man who gave her comfort when her mother was dying, and she was overwhelmed at all she had to do. He then disappeared into thin air but not before touching her, when she experienced a bolt of positive energy going through her. Others have spoken to me about direct experience of angelic beings appearing in their hour of need. They seem to come in a variety of shapes and sizes, depending on who is interacting with them.

There are many people who work with angels in different ways to enhance and help humanity. Angelic guidance can be particularly useful, as someone calling themselves 'Primal' told me, of her own experience,

> "I have personally witnessed two types [of angel] ... the first is the translucent beings of light ... the other was a person that came to help when I was a small child ... Never seen his face, the two times I met him I was sitting in the street on a sunny day. The sun was always above his head and behind it, so I could never clearly make out his face. The first time he showed up it was to give me advice, and the second time was to see how that advice worked.
>
> "The being of light that is on my wall ... I had to look at the nothingness surrounding the being and not focus directly at an area like its face or chest, because wherever I focused my vision the being would instantly start to fade into nothing in that area,

---

[26] *Luke 1:30-32*. Holy Bible, King James Version.
[27] See 'George Washington's Vision and Prophecy About America' recorded in the Library of Congress. https:// sign.org/articles/george-washingtons-vision-and-prophecy-about-america

and it would spread from there very quickly all over. So, I quickly learned to focus on the nothingness ... something I had already trained myself to do."[28]

Angels may be here to help humanity. They can appear in visions or when someone is about to pass over into spirit.

The Religious View of Angels

The religious among us may have different perspectives on angels but all believe them to be created by God. They are an earlier and separate order of creation than humanity, created before man to do God's work and help as his messengers. Buddhism and Hinduism, which follow a different religious path to both incarnation and enlightenment, do not have angels as such but a rich tradition of other spirits.

- **Christian Angels**

    The Christian Holy scriptures teach us there are millions of angels with different attributes and roles, in their mission to help humanity. The Catholic Church has a hierarchy of Angels. The first sphere of nine, the heavenly servants of God, are: Seraphim, Cherubim, Thrones, Dominions, Virtues, Powers, Principalities, Archangels and Angels. The order was, however, mainly designed for philosophical and political reasons, made to resemble the various levels of servants to kings and emperors. The Archangels, those closest to God, are said to have different specialities. The four main Archangels, whom we know by name, also govern the seasons of the year: Michael, the protector (autumn); Gabriel, the messenger (winter); Raphael the healer (spring); Uriel, shines the light of God (summer).

- **Muslim Angels**

    Although we tend to think of angels as playing an important role in Christianity and Judaism, a belief in God's angels is one of the six fundamental articles of Islamic faith. There are many verses and sayings of the prophets in the Quran (Koran) about angels. They are created by God above man and above all other spirits (al-jinn) that wander the Earth. Whilst they are generally seen as beings of light or winged messengers, their angelic function seems slightly more

---

[28] Primal, personal communication, 2016.

judgemental than those of Christianity. As I understand it, Muslims believe that angels record the deeds of man on earth, both good and bad. When a person's spirit leaves the physical word, angels receive the soul on its journey to the afterlife. Each Muslim is said to have two guardian angels, one for day and one for night; both watch over a person and record their deeds before final judgement.

- **Mormon Angels**
  There is a strong belief in angels in the Mormon religion, which again seems to have a slightly different view of their nature than Christianity. Mormons maintain there are two kinds of beings in Heaven called angels. There are those in spirit and those with bodies of flesh and bone. Angels in spirit form have not yet obtained a physical body, or they once had a mortal body and are awaiting resurrection. Angels who have bodies of flesh and bone – sometimes known as Earth Angels – are usually resurrected (reincarnated) beings. For instance, the angel whose figure blowing a trumpet sits atop many Mormon temples, is said to be the Earth Angel that returned to speak to Joseph Smith (from 1823) and helped to inaugurate the Mormon religion.

Other Types of Angels
There are other kinds of angels, including dark angels, in our midst.

- **Guardian Angels**
  There is a belief that we all have a guardian angel, appointed to us from the moment of conception. Some people believe we have more than one. Whether or not we believe in them, our guardian angels are always with us. They have the capacity to protect, guide and advise us, whether we are consciously aware of them or not. It is understood they remain by our side until our spirit passes over, at the moment of death. Some people, through intuition, prayer or meditation, are able to interact and communicate with their guardian angel. They are not only here for our personal good, they can enrich our lives and help us work towards personal and global transformation. Guardian angels can also be assigned to protect a particular group or country, given special commissions at times of crisis or war.

- **Earth Angels**
  Angels sometimes seem to appear to those in need, looking like everyday people, although in retrospect they often have a luminescence or otherworldly glow about them. Historically, angels that interact with humans, such as Abraham in the Christian Bible, came in human form and were only recognised as angels after the event. There are also stories about strangers helping individuals, then vanishing into thin air. Who or what could these beings really be? Some would be actual angels, but many think they belong to a different category. After all, if angels are God's messengers, then it is possible He may choose a human being to perform a particular service or mission.

  There are other alternative views on the true nature of Earth Angels. For instance, it has been said they are spirit guides in training, or those sent back to Earth in order to learn and heal others. This type of angel may have returned to Earth in human form as an angelic being to help others. They could also be visitors in disguise from beyond the stars that have come to Earth to help us. Whatever the truth, I know of people who believe themselves to be Earth Angels, as various websites and Facebook groups can verify, including that of Tracy Anne Morfitt who works with angels and crystals.[29]

There seem to be angelic beings inhabiting spirit bodies, sent from a higher power to help and protect us.

- **Rescue Angels**
  There are many stories of angels appearing in disaster situations or at times of danger, leading people to safety or protecting them from harm or violence. This raises the question, why are some people protected and rescued from a tsunami or a volcanic eruption for instance, and not others? The usual answer relates to God's will – everyone has an appointed time. It may even be to do with karma; we each have to play out our lessons and tests in life. Angels cannot directly interfere, but they can support or nudge, if necessary, especially when we enlist their presence. Our guardian angel can also support us in this way.

---

[29] See Tracy Anne Morfitt's Public Facebook Group, 'Where Do I Fit Into the Jigsaw we Call Life'.

- **Dark Angels**
  There is a view that not all angels work for the forces of light. Dark angels are said to roam the earth seeking out souls into temptation and ruin. While the term "fallen angel" appears neither in the Bible nor in other Abrahamic scriptures, one-third of all angels were cast out of heaven because they defied God and tried to take Him over. Does this suggest that heaven is a physical plane where God is not an overall ruler? Perhaps He allows such a force to exist? The fallen angels followed Satan and joined the forces of evil. Thus, dark angels may be agents of the devil. They are said to encourage selfishness and greed and the acquisition of material goods, rather than kindness and spiritual treasures. They can even pretend to be angels of the light to deceive and entrap us. This belief correlates with the concepts of good and evil, Heaven and Hell, discussed in Chapter 5. To those who can see angels, dark angels seem to look like ordinary everyday angels, but they are neither loving nor offer compassion and healing. They may not show themselves but just communicate with you in your head, leading you telepathically to make bad decisions.

There may also be dark angels. If so, a dark side of reality must exist.

The question is, how can you tell if the beings you are communicating with are messengers from God or dark angels trying to tempt you onto the wrong path? If you have protected yourself from evil before making a connection, there is less likely to be a problem. There are said to be some obvious giveaways though, so be aware:

- Angels have no free will and should not be worshipped in their own right.
- If they require worship or adulation, they are not pure angels.
- If angels offer to do something for you that is not in your best interest, or may cause harm to others, this is not a good sign.
- If an angel offers you earthly wealth or advancement, beware.

The view of dark angels working to tempt humanity onto the wrong path, does give clues to a higher reality. Could the biblical Heaven and Hell as places of good and evil actually exist? Does it mean that a creator God and an evil counterpart belong to the same reality?

Where Do Angels Come From?

Some alternative thinkers believe angels to be a separate branch of Earth evolution. Others believe they are a benign race of aliens that are here to help protect humanity.

Angels may be a separate evolutionary branch of alien beings, keeping us safe as part of an intergalactic mission that we are unaware of.

Belief in angels is not just a matter of religious faith. There are several other perspectives on the nature of their existence, both religious and secular. Here are some varied views on what angels may represent:

- Messengers and helpers from a creator God. Some people understand angels to be His thoughts or the creation of the Divine Mother.
- A collection of beings responsible for the organisation of the inhabited universe, helping mankind to promote harmony.
- A separate species co-existing with us on Earth, with vibrational differences that our senses and technology cannot usually detect.
- An alien race here to help mankind for mutually beneficial purposes.
- A part of one unified consciousness – our ultimate reality.

Of all the definitions and views on angels, the one which most resonates with me is described in the book *Ask your Angels* (1992), by Alma Daniel, Timothy Wyllie and Andrew Ramer. They suggest 'angel' is a generic name for a collective group of beings – citizens of inner space bridging our physical reality with their pure spiritual energy – whose responsibilities include the harmonious organisation of the inhabited universe.

"Angels are intelligent beings, capable of feelings, yet a different species, who have their existence on a slightly finer vibrational frequency from the one to which our physical senses are tuned. This means we cannot perceive them ordinarily with our eyes and ears, but they can perceive us. Our realities interpenetrate one another with their reality encompassing and enfolding ours."

When I asked people what they thought angels are, there were many interesting suggestions. Whilst talking to Sarah Hayward about her work with Angels and Ascended Masters, she gave me a wonderful new insight in response to this question:

> "All that exists within our reality is part of one consciousness. All beings, no matter what their kind, are acting together as one to create a consciousness that creates experiences which increase the knowledge and growth of all."

So, in some ways, any separation between us and them is an illusion. We are all part of the same whole, all with our parts to play. Sarah feels that we cannot possibly know the ultimate truth while we are merely human. When we have passed beyond the veil, all may become clearer.

Angels, of course, may not have come from a creator God, but may be multidimensional beings that are here to interact with mankind for mutual benefit or their own purposes. If so, why are they helping humanity? This makes sense if we are unknowingly part of an intergalactic organisation; keeping the Earth safe may be an important part of a bigger picture. Whatever they are, most people feel that we are blessed to have angels here with us.

Angels may be part of a collective responsible for the inhabited universe, existing in the greater reality that encompasses our cosmos.

Spirit Guides

These are entities from the spirit world that have chosen to look over us and help us in our Earthly incarnation. Their role is slightly different from that of the angels, although some angels do seem to act as spirit guides. Spirit guides are more aware of your life's purpose, and will, if possible, help you to fulfil your aims and desires. They may come from different sources and all of us may have more than one spirit guide, each with a specific role to play. Sometimes the same guide is with you throughout your entire life, although some guides may just appear for a short time to help with a particular problem. They may guide you spiritually and help you through difficult life choices. Some people do manage to communicate with their guides directly. More commonly, they appear in dreams, send messages in subtle ways, or simply feed ideas into your mind.

We hear of those who find their spirit guide sometimes appears to them as a native American Indian or an Eastern guru, for example. When they present themselves in a certain form, does that mean the person has a past life resonance?

A friend of mine had an interesting experience when she visited a chapel in Dorset and sensed a past life connection. A few weeks later, a psychic artist painted the spirit guide that he could see surrounding her aura.[30] The portrait was of a nun, possibly an abbess, yet he knew nothing of her recent experience. This raises the question of whether we are sometimes drawn to places that allow us to connect with one or more of our spirit guides? Perhaps we are, at some level, our own ancestor?

Whoever they are, many of those working with spirit say it is important to create a relationship with our guides, not simply take them for granted. It is reassuring to know there is always someone available for advice or guidance, whether we are aware of them or not.

Entities who are now in spirit may be available as guides to help with our life's purpose.

Sprit guides may be one of several things:

- A friend or loved one who has passed over. (I often feel my late mother giving me advice!)
- An ancestor you never met but is still looking out for you.
- An animal spirit guide. (My husband feels that his last dog, Brutus, is always there for him.)
- A guardian angel assigned to you from birth.
- An angelic being or archangel.
- An ascended master.
- From the fairy or elemental realms.
- An aspect of ourselves from a former life.

Some people also believe that interdimensional beings or those from other planets, from where they may have previously incarnated, also act as spirit guides. Whoever they are, life would be much harder without them!

---

[30] Patrick Gamble, psychic artist and visionary. Find him on Facebook or see www.patrickgamble.co.uk

Ascended Masters

There are other beings present in our reality with us, helping us on our spiritual path. They have usually, but not always, incarnated in our physical world at some time. These are the Ascended Masters, believed to be spiritually transformed, enlightened beings, who in past incarnations were ordinary human beings. They have chosen to come back to the Earth sphere in order to help others along their own paths. In other words, they serve as teachers of mankind from the realms of spirit.

It is said that the Masters can either return as babies or inhabit existing bodies as a 'walk in', a concept popularised by Ruth Montgomery in her book, *Strangers Among Us – Enlightened Beings from a World to Come* (1979). At such an evolved level they can create a temporary physical body to inhabit or incarnate an aspect of themselves as a newborn. They do this so they can serve humanity. Each Ascended Master has a specialisation in one or more areas, so you can call on one with the talents you are wishing to develop. In this respect, they can also act as spirit guides.

In her classic book, *The Cosmic Doctrine* (1966), occultist Dion Fortune claimed to have received communications from the Ascended Masters on the inner planes concerning the creation of the universe and the evolution of humanity. Her philosophy had a significant impact on the nature of reality from the early 20th century.

Interestingly, some of the archangels, such as Michael, are thought, in some circles, to be Ascended Masters that have taken on angelic form. Controversially, Jesus, Buddha, Quan Yin, Kali and Krishna are seen by some as Ascended Masters or prophets rather than gods. On a personal note, I have been told that Metatron, Merlin and Mary, among other Masters, are helping to look after me!

There are advanced beings known as the Ascended Masters who choose to reincarnate to help others along their spiritual path, and for the good of humanity.

The Great White Brotherhood

This is a fellowship or association of Ascended Masters and 'spiritually awake' souls that oversee our Earthly lives. Having passed into the realms of spirit, they work individually and collectively to assist other souls in their Earthly incarnation. Telepathy and sometimes clairvoyance, both considered gifts of spirit, are used to communicate with humanity, regardless of colour, race or creed. Referred to as 'white', meaning 'pure' or 'holy', they are also fighting the forces of darkness.

Some people claim to receive messages psychically from the Great White Brotherhood, a term first introduced by esotericist Alice A. Bailey in the late-19th century. Further developed by Theosophists, including Helena P. Blavatsky, her *Isis Unveiled* (1877) reveals the teachings of a secret group of initiates and Tibetan Masters. It has, however, come to be used as a generic term for any group of enlightened adepts, with the aim of helping the spiritual development of both the living and the dead. The Lucis Trust, founded by Alice Bailey in 1922, continues today with 'World Goodwill', and encourage meditating on a star triangle to visualise light and love pouring into human minds and hearts.[31]

The Great White Brotherhood work with spirit to help souls in their Earthly incarnations and to protect humanity.

Orbs

Orbs are strange, mysterious balls of light which seem to float through our reality. They can manifest in a myriad of different colours and sizes, some clear, others opaque, and some even appear to have faces in them. Opinion differs as to what orbs actually are and where they come from. The general consensus seems to be that they are usually, but not always, balls of energy that come from the spirit realms or other dimensions.

In my research, I found lots of photographs of these mysterious orbs, of all colours and types, taken by very many people. The interesting thing is that not only do they sometimes appear even when the photographer is unaware of them, but on occasion the observer may see them before a photograph is taken.

A surprising number of people admit to orb encounters, including my own family and friends, for example:

C said that one day, in a bungalow where she lived in many years ago, she saw a stream of white orbs travelling through an area of the ceiling and swirling around, then leaving through another

---

[31] See www.lucistrust.org The Lucis Trust have republished many of Alice A. Bailey's books, which include: *The Consciousness of the Atom* (1922), *A Treatise on Cosmic Fire* (1925), and *The Soul and Its Mechanism* (1930).

space. At the same time, she heard "chattering" sounds and had the impression they were pure energy.

G woke up one morning to see there was a large, aqua-coloured orb in the corner of her bedroom ceiling. It lit her way down the hall to the bathroom!

P who is healer, said she saw orbs of many colours all the time when she was giving healing to clients.

X felt that the spirit of her departed twin sister was in an orb, which followed her around sometimes.

M saw orbs when he was a very small boy, with something inside them. Now older, my grandson does not remember.

T saw gold-rimmed orbs swirling around in a haunted ballroom.

I remember first seeing orbs without realising what they were. At the time I was in great distress, when balls of emerald green light, which I know now to be healing, would often appear.

Why don't we all see orbs? What we see depends on the quality and sensitivity of our sense organs, our eyes in this case, our perceptions determined by how our brain organises the information it receives. So, there are several things going on here, which varies from person to person. There is much more stimulus around us than our eyes can see and our brain can process. It also acts as a filter and sometimes simply cuts out the things it doesn't recognise. Those people involved in energy or spiritual work are obviously more sensitive to seeing such things, beyond what is sometimes put down to an optical phenomenon resulting from 'back-scatter'. They often have an enhanced inner vision.

As Christian Kyriacou tells us in *The House Whisperer* (2014):

"As the veils between the worlds of physical and metaphysical dimensions are becoming more permeable, we are gaining greater awareness of unseen energies. Their interaction with human consciousness plays a large part in the manner of how their presence is revealed."

My initial research reveals at least four different types of orbs, although some may fall into more than one category.

- **Coloured orbs**

  These are said to appear when needed and are often connected with healing or protection. They frequently appear in the elemental world of nature. Different coloured orbs seem to fulfil different purposes for different individuals. As ever, there is often a slight variation between practitioners as to what the various colours mean. Many people begin to notice their own 'code', as it were.

  Here is a general list of my own colours and their purposes:

  Blue (dark): Shy, spirit.
  Blue (light): Tranquil, peace.
  Blue (medium): Protection and comfort.
  Brown: Earthbound or danger.
  Gold: Angelic, unconditional love.
  Yellow: Solar energy or sometimes caution.
  Green: Healing or spiritual.
  Lavender: Messenger from God.
  Red: Active, busy, anger, may also offer protection.

  There may be more variations and colours out there which I have yet to come across. My own experience of seeing green orbs happened on the occasions when I needed it. I have also heard black or brown ones sometimes referred to as malevolent, or to warn of danger.

- **Spirit orbs**

  Some orbs are thought to be ghosts or spirits coming to Earth in the form of light. Tracy Anne Morfitt told me she believes they are trying to communicate with the living human world once again.[32] Some may be trying to communicate with certain individuals or loved ones. These are often, but not always, clear or white. It is said to be easier to come back in orb form than to manifest in a spirit body.

  Spirit orbs have also been experienced at ancient sites, such as

---

[32] Tracy Anne Morfitt, personal communication, 2018.

barrows and stone circles. It is as if the ancestors are trying to communicate or attract one's attention for some reason. Sometimes they are simply mischievous and playful, as if they are pleased to remain in a much-loved place, perhaps close to their tribal leader or as a guardian of the site.

Larger orbs require more energy. Sometimes what appear to be different life-forms are visible inside them. Others may just be passing through our reality. Streams of these kinds of orbs are often seen travelling in groups. Those seen by C may fall into this category, and she did hear a chatter of communication between them. This raises the question that if spirits, in orb form, are floating around the greater cosmos, what does this tell us about our reality? Is it possible that our souls float around after death and/or before incarnating in physical form? This would mean we do go on in some form after death, and there is a place or dimension for us to go.

There are other occasions when orbs can appear as if to join in the party! I have witnessed many orbs appear in photographs above a person's head, surrounding the stage on which they have just given a presentation. It is as if the content of their talk and the reaction of the audience has created a heightened atmosphere. Others would say it is simply the spirits in the building responding.

- **Dark orbs**
  These are the kind of orbs that tend to give us a heavy or uncomfortable feeling. They are often to be found in damp basements, abandoned rooms and derelict buildings. Spaces which are rarely used, full of clutter or neglected, can accumulate a concentration of stagnant energies owing to lack of air and human interaction. This can allow darker energies to creep in, which may appear in the form of many small orbs that can feel somewhat malevolent. These orbs rarely whizz around or interact. They can also manifest in places with a sad history and feed-off unhappy emotion.

- **Alien orbs**
  These are perhaps the most contentious and denied form of orbs. The multi-coloured ones are sometimes said to be alien orbs. They could also be alien probes. There have been reports of them being seen around crop circles, emerging from UFOs, and even a suspicion they may be scouting out possible abductees. Denial by everyone in

authority is, of course, the norm. Alien interaction with humanity has been long documented (see Chapter 7). Alternatively, are these orbs nothing more than advanced energy vehicles?

The very existence of all types of orbs gives us possible Super Clues to the nature of our reality. It is known that orbs interact with etheric space and can communicate with humans, at some level. What they are and where they come from is a matter of much conjecture. However, it is believed that these light structures are energy beings existing within the quantum field.

It is possible that departed souls can revisit our reality as spirits in orb form. Orbs can manifest at sacred sites, places of heightened energies and be an indication of stagnant energy. There may even be alien orbs.

What do the Super Clues embedded in the existence of the spiritual and angelic realms and orbs say about our reality?

Summary
The existence of angels and their apparent mission to help and protect humanity, throws up several different reality scenarios. The angelic realms show us a more ordered and hierarchical existence, giving us a picture of a universe above and beyond our physical Earth plane. This hints at a higher power for whom they may be working, perhaps a God or a designer of some sort, or even an alien race. It also indicates a deliberation about the creation of the universe and, once again, a greater overall consciousness than we are normally aware of.

Whilst there are spirits working against us and there are also believed to be dark angels, beings in the form of angels and spirit guides can be viewed almost as a missing link, giving us glimpses of an understanding of the energies that flow through our universe. They are available to help and guide us throughout our Earthly incarnation, and generally bring some form of protection against the forces of evil, whether we are aware of them or not.

Orbs only seem to have appeared in our reality in more recent years, some say with the advent of the digital camera. If orbs can manifest in both a healing sense and be of a more sinister or warning nature, then where do they come from? Are they inherent in the structure of the universe, or are they being sent by some kind of external agency? If orbs are alien visitors rather than from the angelic realms, or simply departed spirits, then this again indicates a more complex and directed, possibly multidimensional, universe than we are taught exists.

What this all means, however, is that even our physical reality is more interfered with, supervised and monitored than we may realise. As incarnated human beings we are co-creators in a greater dimensional reality, both physically and spiritually.

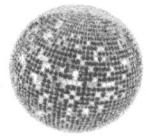

CHAPTER 9

Super Clues embedded in Nature Spirits and Elemental Beings

BEFORE I BEGIN, WE HAVE to ask the fundamental and obvious question, if the elemental realms are all around us, as with all spiritual beings, why can't we all see them? Well, as discussed previously, there are at least two possible answers.

Firstly, as seemingly solid physical beings, we are all basically made up, at the atomic level anyway, from little more than vibrating energy. This energy vibrates at certain frequencies. Our sense organs are tuned into those frequencies. Other beings vibrating at different frequencies we may simply not detect, although they may be all around us. We may, however, be able to communicate with them in our minds.

Secondly, all we perceive through our sense organs is co-ordinated by our brains, which make sense of things and sees what it expects to see. This also applies to our perception beyond what is visible to the human eye. Even if other beings are there, our brains may not recognise them as valid images, so not register their existence. The truth is, that some people do have the facility for seeing and/or sensing these otherworldly beings and are sometimes able to work consciously with them.

The interesting thing about exploring other beings that share our reality with us, is the light it shines on the big picture of our existence. What roles do the elementals and nature spirits play in the jigsaw of our universe? What does their existence mean for the true picture of our reality? When you start looking, you find many more clues out there than would have been thought possible.

Our Earthly reality is teeming with spirits of all kinds, invisible to most people. There are those that help humanity to maintain our environment, those spirits that work with departed souls, and those which simply pursue their own ends and may even work against us. There are also ghosts, poltergeists, devils and other disembodied spirits, as we encountered in Chapter 5.

Every culture has its own names and characteristics for different types of spirits. For instance, Arabic and Islamic folklore have teachings which assume that spirit beings occupy a parallel world to mankind. They are the djinn. Like humans, they can be good, evil or neutrally benevolent. According to the Quran, the djinn, literally meaning "hidden from sight", were created by Allah from a smokeless fire. We are familiar with them in European folklore under their Anglicised name of Genie. This was the name of the being in the children's story that was trapped in a lamp and released by Aladdin, who then obtained three wishes as a reward!

Many of these beings found in myth and legend have been known and talked about since the beginning of civilisation. As mentioned, some people can see and interact with them, although most cannot. They have a higher and finer vibration than humans, which our sense organs cannot always cope with. Some forms can show themselves to us if they choose to, although we do need a certain level of sensitivity to recognise and interact with them.

Another class of being considered integral to our reality are the nature spirits (devas) and elemental beings, which are said to be essential for maintenance of our biosphere. There are also the fairy folk and little people. Some, but not all, help the environment. Some may even be hostile to humanity. There are two main difficulties in understanding these beings. Firstly, because there are many different and confusing names for the same type of being and fundamental differences in how people classify them. Secondly, there is problem with separating fact from fiction. This is largely because there is a huge gaming industry that create fantasy characters using the names of existing beings and make up others. Fact and fiction get intertwined and blurred, such that game 'facts' and labels become the surreal identities of reality.

Nature Spirits
There are said to be thousands of different types of nature spirits inhabiting our reality, doing different tasks to help maintain life on Earth. It is said that we could not exist without them. As with the angels, they may just be another branch of Earth evolution developing alongside humanity, or they may have been placed here by a designer God.

SUPERCLUES EMBEDDED IN NATURE SPIRITS AND ELEMENTAL BEINGS

In this section, I will be looking mainly at the devas and fairies.

- **Devas**

    These first appeared in early Vedic literature as a class of divine, benevolent supernatural being. The term 'deva' means *shining one* in Sanskrit. In Buddhism, it refers to different types of non-humans who share certain god-like characteristics, including very long life and strength.

    When one begins to research what a deva is understood to be today, the matter becomes even more complex. The term is often used to describe a multitude of beings made up of etheric matter or energy. Some people use the word 'deva' to encompass all nature spirits, or any of the spiritual forces and beings that inhabit nature. This also seems to include the fairy folk. Others propose they are a type of angel, or at least in partnership with them, looking after nature and the Earth rather than its people.

    Devas are considered magical or supernatural beings with a higher vibrational frequency than humans. They are only visible to some sensitive individuals, especially those with an activated third eye. Also known as the mind's eye and related to the pineal gland, this chakra, situated at the centre of the forehead, allows for inner vision.

    Devas are connected to the earth and the green environment. They are thought to direct and be responsible for the health of our ecosystem. One example of the benefits of working positively with devas and nature is the Findhorn Foundation. Peter and Eileen Caddy and Dorothy Maclean, found themselves in dire straits, living on a caravan site at Findhorn, on the east coast of Scotland. By working with the nature spirits and devas they were able to grow plentiful vegetables, which should not otherwise have been possible on such poor, sandy soil. Findhorn grew into a large community, and became a foundation dedicated to teaching the principles of co-creating with nature.[33]

There is a class of being, made of spiritual energy, called Devas or nature spirits, that are here to help the biosphere survive and thrive.

---

[33] The Findhorn Foundation also run events and workshops. www.findhorn.org

- **Fairies**
  When people refer to fairies, they often mean the diminutive magical folk that inhabit our reality. Although considered devas as they are nature spirits, fairies also have magical powers. The terms deva and fairy tend to be used interchangeably but, as I understand it, not all classes of fairy folk are devas and not all deva are fairies. With subtle differences, the edges can get very blurred. All the "little people," such as pixies and leprechauns, are considered fairy folk. The beings we know as traditional fairies are only part of the larger magical fairy family.

  When one mentions fairies, most people imagine the enchanting pictures in our story books. Small and pretty, like little people with small wings, as popularised by Cicely Mary Barker in her series of books on *Flower Fairies*.[34] In fact, most people who can see them say they see 'pinches' of vibrating energy. Whilst fairies can choose to show themselves in the traditional story book form, they can appear in different guises as our brains automatically perceive them in a form we expect to see – as, of course, with all these imperceptible energy beings.

  As part of the family of nature spirits or devas, fairies too are here to help maintain the biosphere. In their nature spirit role, they are said to be an essential component in maintaining the health of the planet. Like the angels, fairies are fundamentally protecting us in our environment, but they are much closer to the physical world.

  Fairies also seem to have magical or manifesting powers, and traditionally interact with humanity – although this is not always in a positive way! They are said to have an excitable energy and can be very mischievous. They can either help or take against and harm, individual humans.

  In her book *The Real World of Fairies* (1999), Dora Van Gelder Kunz gives a wonderful description of fairies being creatures of pure emotion and sensation. Interestingly, the author suggests they also establish relationships with other fairies, with plants and animals, as well as with humans, by adapting their vibration to that of the being they wish to relate to.

---

[34] The first of Cicely Mary Barker's Flower Fairy series, with botanically accurate illustrations, was *Flower Fairies of the Spring*, 1923. See www.flowerfairies.com

Fairy characteristics are hard to pin down. Each person seems to have a different viewpoint, experience or understanding of what they are and their purpose. This list covers some of these optional 'Fairy Facts':

- Every plant is said to have its own fairy helping it to grow.
- Flower fairies are said to wear the colour of the flower they protect, or hats made from its petals.
- It is possible to communicate and work with fairy energy.
- They can help humans to harness the Earth and Moon energies to manifest, especially in the garden.
- Offerings were traditionally left for the fairies to enable your crops to grow, attract wealth and protect you from harm.
- They love to dance and sing.
- Mostly friendly, fairies can at times get mischievous and play havoc with humans, especially when annoyed by them.
- They can gift you presents or manifest magical powers and skills for you.
- They live alone or in colonies; there may even be fairy houses or settlements.
- There is a belief that some fairies may be fallen angels.
- In fairy stories things often happen in threes and sevens.
- There are spells on the Internet for summoning fairies.

Fairies help the biosphere. They can interact with and manifest for humans. Their energies can also be used as a human manifesting tool.

What insights into the nature of reality does the very existence of nature spirits, devas and fairies give us? This comes back to the fundamental question of existence. Are fairies, like the angels, put here by a creator God or intelligent designer? Or are they just part of an evolutionary process, and have developed alongside humanity largely unseen?

Other Fairy Folk
There are many other "little people" in the fairy kingdoms, as well as beings such as elves and imps, said to routinely interact with humans in our reality. These include brownies, leprechauns, dwarves, sprites, merman and mermaids, asrai (aquatic fairy), selkies, pixies, and many

more. There are too many to describe or mention. All seem to be considered fairies. Although not all are nature spirits, they are said to have magical powers, but the way they interact with humans and the environment seems totally different. They may be useful and kind, or evil and nasty. Some are even purported to have intimate relations and breed with humans, creating monstrous children. They can carry off children to fairyland, leaving changeling substitutes behind. Plants and herbs such as St John's wort and yarrow are potent against unwanted fairies. The protective kind are often associated with the household hearth and commune with nature, in more recent times helping to look after gardens. Female fairies may tell fortunes and prophesy births and deaths, especially if you believe in them!

Whether all these categories are real or fictitious is almost impossible to discern. Many are mentioned in songs and oral tradition. The term 'fairy' only, however, appears to go back to the Middle Ages in Europe. It was the German brothers, Jacob and Wilhelm Grimm who collected and published their compilation, *The Original Folk and Fairy Tales*, in 1812, and the Danish writer Hans Christian Andersen whose enchanting *Classic Fairy Tales* include 'The Snow Queen', 'The Ugly Duckling' and 'The Little Mermaid', that first popularised fairy stories for children and adults alike. Following a similar pattern that begins "once upon a time," the hero or heroine usually meets with a villain: a witch, ogre or monster. Magical characters come alive on an archetypal journey that often features a king, queen or a castle, and ends with the triumph of good over adversity.

Here are descriptions of just four kinds of fairy folk.

- **Leprechauns**
  Perhaps the best known in British culture, apart from traditional fairies, are the beings known as leprechauns. Usually dressed in green with pointed hats, which were traditionally red, an entire leprechaun industry has evolved around the myths and legends of these little men. While classified as fairies, unlike some fairies they don't seem to be considered nature spirits, but they too have magical powers. Leprechauns are said to be descendants of a group of magical beings that served under the Gaelic goddess Danu, long before humans inhabited Ireland. Unlike the sweet flower fairies, Irish fairies are legendary – they can be nasty, capricious creatures, whose magic might delight you one day and kill you the next! They

often need to be appeased to stay on their good side. Known for their abilities as cobblers and for their hoarding of gold, this can usually be found at the end of the rainbow. Today, they seem more mischievous than malevolent. Should a leprechaun be caught, in order to escape it may grant the protagonist three wishes or proffer gold.

- **Goblins**
  These creatures, found mainly in European stories from the Middle Ages, are usually monstrous. What goblins look like and what they do depends on their country of origin. They are always depicted as small, grotesque and greedy, especially for gold and jewellery. Like elves, the problem for any serious researcher is that goblins are heavily fictionalised in gaming videos and fantasy books, where they are usually the bad guys. J. R. R. Tolkein in *The Hobbit* (1937), describes the ugly goblins who lived under Misty Mountain as "cruel, wicked and bad-hearted." It is hobgoblins that often have magical abilities and can be mischievous, such as the clever and quick-witted Puck in Shakespeare's *A Midsummer Night's Dream*, who helps his master play a trick on the fairy queen.

- **Imps**
  Often described as mischievous rather than seriously threatening, and as lesser rather than more important supernatural beings, the attendants of the devil are sometimes described as imps. They are not dissimilar to mischievous fairies and are usually of small stature, lively and prone to playing pranks. Imps are sometimes known as wild tricksters, uncontrollable and fun loving – impish even!

- **Elves**
  Elves are diminutive shapeshifters who may have sprung from early Norse mythology. English male elves are depicted looking rather like little old men, while elf maidens are invariably young and beautiful, capable of seduction. Like men of their time, elves live in kingdoms found in forests, meadows, or hollowed-out tree trunks. They developed a reputation for mischief and are said to sometimes cause illness in livestock and take terrible revenge on humans who offend them. The most famous of these magical beings are the elves that work for Santa Claus at the North Pole.

There are fairy folk living among us that may be malevolent beings working against humanity for their own ends. Others are said to have magical powers and are here to help and support us.

Elemental Beings

An elemental being is a type of magical entity who personifies an elemental force of nature, and control the natural powers derived from their own element. The elementals create and help life on Earth to exist and flourish. In antiquity, there was said to be four elements which made up the Earth: earth, fire, air and water. Thus, our understanding of these elemental beings is based on their origins. The Chinese have iron as a fifth element, and there are many other cultural variations.

Elemental beings, on one level at least, are said to be made of only one element. There is a corresponding type of being for each of the four elements, as defined by the 16th century alchemist, Paracelsus. The Gnomes (earth), Sylphs (air), Undines (water) and Salamanders (fire), all work hard to sustain life on Earth on our behalf. We are, unknowingly, dependant on their selfless service. They can each move through their own element, just as humans can move through air. As with all magical beings, we cannot always see them, although those with a special sensitivity may be able to do so. Whilst they work hard for the planet, and are often ignored and taken for granted, mankind may irritate them. Yet their qualities and sense of wonder and beauty in nature offer great gifts for humanity.

There are myths and legends concerning these special, vital elemental beings throughout the ages. Precise information is, however, difficult to obtain as sources do not always agree. This is compounded by the fact that, as with all nature spirits and legendary magical beings, the rise of the fantasy game industry has blurred even further their roles and qualities, which has confused the information stream. Other than for the serious student of mythology, separating fact from modern fiction is harder than ever. I have distilled below the available information as accurately as possible.

- **Gnomes (Earth)**
  Dumpy and grumpy, gnomes guard the mines and treasures of the Earth. These are the nature spirits who serve at the physical level. Popular as garden ornaments and usually portrayed as wizened old men with pointed hats and beards, they are said to move through

the earth at will, where they live in the interior. There are billions of gnomes that work tirelessly all year round to make sure all living things get whatever they need. Also associated with helping on the farm, they process waste on land and mitigate, wherever possible, pollutants and negative energies, as well as those toxins poisonous to the Earth and humanity.

- **Sylphs (Air)**
  Sylphs are believed to be an offshoot branch of the Sidhe, a supernatural race that live in mounds or hills, comparable to fairies or elves. As graceful beings of the air, sometimes they resemble clouds. Mostly invisible, at other times seen as tall, thin humans with large wings. Sylphs can soar above the Earth at will and travel great distances. They can be mistaken for angels, or angels can sometimes be confused for sylphs. They are said to purify not only the air, but also the mental plane. Sylphs are invaluable to our planet, especially with so much modern-day pollution and chemtrails. There are also enchanting woodland sylphs, usually male, sometimes confused with the tree nymphs known as 'dryads' from the Greek *drys*, meaning oak.

- **Undines (Water)**
  Undines are supposedly the spirits of water; some say present in every drop. The undines protect water. They also purify waters that have been poisoned by sewage, industrial waste, chemicals, pesticides, and other harmful substances. They work tirelessly to cleanse the polluted seas. Undines usually, but not always, resemble female humans.

  Mermaids are considered a sub-species of undines, but there are others. There also exist merman, both in human form from the waist up and fish-like from the waist down. While mermaids interact with merman, they may also interact with male humans. Mermaids have a reputation for seducing sailors into the depths, whilst mermen can supposedly foretell disaster and incite shipwreck and storm. So, are they together in league against men, or could an aspect of mermen be protectors of seamen at times of need from alluring mermaids? There are also sprites, which although often associated with the fairy realms, are related to electrical discharges that can result in coloured images in the night sky, and to visions of spectral ships.

- **Salamanders (Fire)**
  Great claims are made for salamanders, the elementals responsible for controlling the element of fire. They are said to appear as slender adult male humans or sometimes look like lizards, made from flames. They infuse our reality with the creative energies necessary to sustain life on Earth. Capable of wielding both the most intense physical fires and the purifying fires of spirit, they control the spiritual aspect of the material oscillation of light within the nucleus of every atom. Salamanders are agents for the transfer of the fires of the subtle world for our everyday use. They can absorb negativity from humanity and are also capable of putting out fire. Unlike other elementals, there exist many recognised species of amphibious salamanders.

It has been brought to my attention that there may be other elemental beings, who play a vital role in our reality as both co-creators and protectors of humanity. Storm, as explained to me, may be one of these, an enhancement element that freely combines with the other four elements to create earthquakes, tornadoes, hurricanes, infernos and tsunamis.

Elemental beings may be fundamental to our creation. They create, control and maintain the natural powers of the Earth, as derived from their own element.

This gives just one layer of elemental beings. Very excitingly, there are said to be more elemental beings than are usually recognised, fundamental to the creation of our reality. Some people believe that the elementals created the world from their own realms. In delving into this world of nature beings and elementals, I have been shown a totally different and magical way of looking at our creation and existence.

Tanis Helliwell's book *Summer with the Leprechauns* (2011), is a true story of the author's experience of living and interacting with these seldom seen beings who shared her cottage in Ireland. They taught her about the evolution and importance of human interaction with the fairy folk and the elementals, for the benefit of both our races and the Earth.

Slovenian born Marko Pogačnik encourages us to meditate and talk to these beings and to re-establish the connections between mankind and the powerful realms of the Earth. He describes, in *Nature Spirits &*

Elemental Beings (1997), not only his own experiences but their roles in the web of life including trees and landscape energies.

The following list of other elemental beings was given to me by a friend, Matthew Gibson, in 2018. He reminded me that when I refer to our reality, I mean the second level astral realm where we experience physical life. As well as the four main elements: earth, fire, air and water, there is a fifth element, aether – the element of spirit which pervades everything. He states there are five further elemental forces:

1. **Life**: the element that gives us life, the life-force which science cannot yet properly explain.
2. **Light**: the elemental consisting of photons within the electromagnetic spectrum; perceived by our eyes it allows us to see.
3. **Onyx**: the elemental generally known as silicate which is essential to the soul.
4. **Death**: a transitional element, vital to transform physical matter into etheric energy and enable spirit and soul to move on to the next plane.
5. **Dark**: the anti-proton elemental, which is behind dark matter and black holes, etc.

It should be mentioned that each of the 11 elements supposedly has a male and female aspect, giving 22 in total. Whilst they occupy their own realms, they do venture into our physical world. As human beings, we are fundamentally made up of these elements. This is a whole new way of looking at our reality. Unprovable but fascinating!

What do the Super Clues embedded in the existence of nature spirits and elemental beings say about our reality?

Summary
If these beings do exist as described, then, once again, mankind is not the independent, self-sufficient race we may believe it to be. The tasks carried out by the nature spirits and elemental beings are essential to a thriving biosphere. The elementals may well be fundamental to our creation and survival. Their purpose in maintaining the Earth to allow us to exist is an indication of deliberate design, which comes through very strongly.

Not forgetting the other races of fairy folk, the "little people," some of whom indeed may be hostile to humanity. They are known across many cultures by different names for similar reasons. They may just be a different, parallel branch of evolution, or part of an overall picture we have yet to understand.

The concepts of good, neutral, or evil are human labels that offer us interesting clues, in this context also to be found in our classic fairy tales. Does this mean the constructs of good and evil are creation-wide, and serve some sort of purpose in our reality? All sources agree that these beings are essential to the Earth and humanity's existence. Whilst in one sense this is a simplified view, glimpses of realms that exist beyond the visible do begin to emerge.

Our nature spirits and elemental beings are a crucial reminder of the levels of creation in the astral realms, signs of wonderment of an even bigger picture of reality.

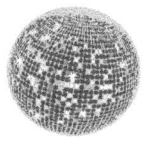

CONCLUSION

IN THIS BOOK, I HAVE identified and explored many Super Clues embedded in various aspects of our universe. These nine chapters encompass our physical reality, human consciousness, science and cosmology, our spirit bodies, earth and universal energies, ancient and modern wisdom, alien beings from other dimensions, the angelic realms, spirit guides and orbs, nature spirits and elemental beings – all of which are part of the big picture of our reality.

They are, as the title says, Super Clues. While not necessarily definitive or complete accounts, they provide insights into the gaps in our present understanding. They are intended as a guide, to offer indications of possibilities and bring a new awareness to some, giving a more comprehensive and multidimensional perspective for those of us who have chosen to incarnate on planet Earth.

Both science and esoteric wisdom are making great strides in furthering our awareness and knowledge of these topics. Advances in the twenty-first century may allow a much greater clarity for humanity in our understanding of who we are and what we are doing here.

As I said at the very beginning, the greater reality we live in is fundamentally unknowable by us, as limited, human beings. We are probably not meant to know all that there is to know. Other more advanced beings, not in human form, may well have seen to that. All we can try to do, in order to comprehend our own reality, is to look at the clues we are able to recognise and try to glean a better idea of what is going on in the greater scheme of things.

Many of these topics are outlined in my first book, *Aspects of Reality – A User's Guide to the Universe*. In this, my second book, I have explored again, in greater depth, my initial topics (or aspects) and expanded my research into new territory. When one looks closely enough, it becomes apparent that the same indications or themes are to be found in several different, seemingly unrelated, areas. This led me to believe that I was beginning to unravel some of the mysteries towards gaining a deeper understanding of the interconnectedness of the nature of ourselves within our universe.

So, what did I find? There are nine main themes (chapters) running through this book, *Super Clues to Reality*, which are then divided into sections. I will attempt to sum up each in turn.

1. **Our Physical Construction**: The key to understanding, at a human level, what we and our universe are made of, is to recognise that we, and everything else around us, is constructed of no more than parcels of vibrating energy. These energies, or quanta, are subject to the laws of quantum mechanics. The apparent solidity and order that we perceive is simply an illusion. The reality, no matter which version of creation we go along with, is that we are energy beings living in an energy matrix. This means not only are we all connected at an energetic and quantum level, but human consciousness can be seen to affect or change the apparent physical reality around us and may even play a part in its creation.

2. **The Nature of Human Consciousness**: This reveals perhaps the most important Super Clue of all. Not only is the reality that we know we are here the only way to be sure we are alive, there is evidence that consciousness is separate and separable from our physical bodies and goes on in some form after physical death. This suggests the possibility that our essence or soul, is transferable. The fact that it has somewhere to go is significant and indicates there must be realms and dimensions in our reality that we can neither see nor detect but must be there. Our existence, and in turn our reality, must surely be part of a much bigger picture than we are presently aware of. The art of manifestation also plays a part in our experience of being human. This raises the important question of why, at a soul level, did we choose to incarnate on planet Earth? There must be an overarching, unfathomable reason.

3. **Big Science and Cosmology**: The physics of the greater universe and the many and varied cosmological theories, including the nature of time, are fascinating. For example, we could be living in a holographic universe. If we are holograms living in multiple universes simultaneously, then where are the original blueprints? Science itself postulates

there are many dimensions in our universe which cannot presently be seen or recognised. We find creation myths worldwide telling us stories of much earlier civilisations, the existence of ancient technologies now lost to us. Some of these advanced technologies are presently being worked on but are being kept quiet about. Different cosmologies also present other theories of alternative origins of creation. Anything is possible, which leaves us no closer to ultimate answers – only clues.

4. **Our Spirit Bodies**: The human body is far more than just a physical entity. It is understood to have a soul and many spirit layers, which are reflected in our own unique aura. Some of these layers can be used to maintain our well-being and for healing. The physical body, with its chakra system, may be connected to a vibrational source beyond our present understanding. The very existence of the spirit body hints at a strange and powerful reality beyond physical life on this planet. As individual members of humanity, we may be fragments or aspects of a collective consciousness, connected to what we call Source.

5. **Life After Death**: In discussing topics such as ghosts and spirits, karma and reincarnation, and the Akashic records, what happens to the soul after death assumes a reality of multiple lives. The possibility that we have lived before is accepted in many cultures. Having our deeds and misdeeds recorded in the Akashic records and a degree of personal choice in our soul contracts, leads to the karmic principle of "you reap as you sow". The concepts of good and evil, heaven and hell – should they not simply be human constructs – assume an overlord directing our lives at a higher level, and may suggest that a conventional God does, after all, exist. There may even be a cosmos-wide battle taking place. All this points to the fact that we may not be as autonomous as, for the most part, we tend to think we are.

6. **Universal and Earth Energies**: To explore how the Earth is dependent on the heliosphere and on solar and lunar forces to allow us to survive and thrive, prompts the observation that somehow all is far too intricate and complex to have happened by accident. If there is a grand plan, then who or what is the designer? The possibility that the Earth and all heavenly bodies are sentient beings is another intriguing prospect. The Gaia principle indicates that humanity, earth and cosmic energies are all interconnected. Many of these energies, including

geopathic stress, can be identified through dowsing. Planetary and alternative grid theories give us clues that there may be portals to other dimensions at places of gravitational anomalies.

7. **Alien Beings and Other Dimensions**: The very existence of alien and extra-terrestrial beings in our reality suggests we are not alone. The evidence of ancient aliens and reports of unidentified flying objects throughout history, indicates the probability that, from the very beginning, we never were. There are clues of creation-wide contact and communication, in which humanity may unknowingly play a part. There is also evidence of lost knowledge, ancient technologies, and even advanced civilisations, in which extra-terrestrials may have been closely involved. It is also thought that our governments have covered up the likelihood of intergalactic intervention. Our reality exists on a bigger stage, with more players and with more to lose, than most of us are aware. We may even be living in an alien computer generated reality.

8. **Spiritual and Angelic Realms and Orbs**: The existence of angels may suggest that a conventional God does, after all, exist. Angels seem to be here to help humanity; the question is why and what purpose do they serve? The real clues lie in their origins and the nature of these beings. Most interesting is the notion that the arbitrary separation we make between us and them is an illusion. We may all be part of an archetypal collective unconscious, co-creating a shared reality. Orbs also permeate our reality, whether we are aware of them or not. Some of these energies, both discernible and invisible, appear to be here to help mankind and the planet, whilst others serve to warn. Some orbs embody spirit beings or alien visitors; they may be joyous or scary and are sometimes simply an indication of stagnant energy. What reality uses orbs as transport or reconnoitre vehicles? This gives us multiple reality clues. On the inner planes, the Ascended Masters and the Great White Brotherhood are helping both the living and the dead. We are certainly not alone, and we are most certainly being watched. Why and how we are being communicated with remains a mystery.

9. **Nature Spirits and Elemental Beings**: To explore other beings that inhabit our reality with us, reveals a reality teeming with a life-force many of us are unprepared for. Some of these beings, such as the nature spirits, may be essential in maintaining our Earth's ecosphere to allow

for human habitation. The existence of nature spirits, devas and fairy folk, often considered make-believe, show a very intricate reality that probably could not have come about by accident. The elementals work tirelessly with our elements of earth, fire, air and water, and may be fundamental to our creation and to the earth as a living entity. Who could have sent these beings to help us? Again, this all points to co-creation, if not the deliberate design of an intelligent mind.

So, what do these Super Clues mean for our reality? There is a physical reality which we are aware of and each perceive differently, and a greater multi-layered reality beyond. What we do know for sure is that our physical world appears, on the surface at least, to be fixed, yet we are nothing more than energy beings vibrating in an energy matrix. We only know we exist because we have been given consciousness. Our soul is the immortal part of each of us which continues beyond death, an aspect of our consciousness that returns to Source. All points to the likelihood that we incarnate in many physical bodies in different cultures and lifetimes.

We are surrounded by many other beings, discernible at subtle levels of perception. There exists in our reality an extended family of beings that inhabit both the natural world and other dimensions which we have yet, or have no need, to interact with. They may be around on our planet, inhabit other worlds in our galaxy, or come from unknown universes. They may or may not be helping our lives and survival. Whatever else we might think, we are not alone on planet Earth.

There may also be layers of existence, such as the astral planes, outside of physical creation. We do, however, see glimpses of a non-physical universe interacting with our own. For instance, spirit or ethereal energy – the life-force – helps to maintain all living things. Other beings, such as the elementals, are essential in maintaining our biosphere. We are a water planet and humans comprise 70% water – our oceans and rivers are dependent on solar and lunar forces, and determine our survival. Surely, all this indicates a planned reality at some level, not just the result of random evolutionary processes.

In a cosmic sense, who or what is responsible for the design and construction of our universe? A creator God or universal intelligence, an intelligent mind, cannot be ruled out. It is also possible that human consciousness may have a part to play in the co-creation of our reality. The concept of a fixed reality looks shakier than one might think!

A final word ...
I began with the questions, who are we and what are we doing here? I can at least now, to some degree, begin to answer both parts to this. We are spiritual beings who have chosen to incarnate on planet Earth. We are here to experience a physical lifetime in human form, to gain experiences so our souls may develop and grow – our ultimate purpose is to connect with Light and Love at the centre of all things.

Good and evil, light and dark, flow through all aspects of reality. These concepts arise repeatedly in my research and led me to the myths and legends of many cultures; their creation stories are remarkably similar worldwide. This duality is woven into the fabric of the universe – the warps and wefts of life – perhaps for the benefit of humankind. Equally, they could be human constructs or simply an illusion. All bring different reality perspectives, and all give us evidence that we are part of much bigger interactive picture than we may be aware of.

In assembling these *Super Clues to Reality*, we are left with a vast conundrum. Our reality is shown to be both multi-layered and multi-dimensional. However, our world is both indeterminate and capable of change on one level, yet very structured and hierarchical, seemingly pre-planned, on the other. Our physical reality is more uncertain than we may like to believe and could well be created as we go. I often describe life as an interactive video game, whereby we set the rules within certain confines and limitations, whilst seeing glimpses of the past transcending the present into the future.

The truth about reality remains largely beyond human comprehension. The complexity and integrated balance of our universe to enable our survival, strongly indicates a deliberate design or master plan, realised by some kind of cosmic force or greater intelligence.

We will need further Super-Super Clues to find the ultimate truth. This is all I can say for now – more insights and research needed!

ABOUT THE AUTHOR

Marian Matthews is a thinker, blogger and writer, who is fascinated with finding out the truth about our reality.

From a very young age, Marian was curious about how the world around her worked. Interested in both science and in esoteric matters, once the children had grown up and left home, she began serious research into who and what, as human beings, we really are and what we are doing here on planet Earth.

Her first book, *Aspects of Reality – A User's Guide to the Universe* was published in 2012. When giving talks about this subject matter, she began to recognise what she saw as Super Clues emerging from the aspects of reality she was exploring, leading to an even bigger reality beyond that which we are normally aware of. This resulted in another seven years of thinking and research which culminated in this, her next book, *Super Clues to Reality*. In sharing her findings, in layman's terms, she appeals to those on the path of discovery.

Marian lives in small town in north Dorset with her husband, Toby. She is an active mother and grandmother, and has many more adventures, including sailing and writing books, planned in the future.

Thoughts on Reflections of Reality
www.marianmatthews.com

BIBLIOGRAPHY

Alighieri, Dante, *The Divine Comedy: The Vision of Paradise, Purgatory and Hell*. Translated by Henry Wadsworth Longfellow (1807-1882). http://www.gutenberg.org/files/8800/8800-h/8800-h.htm

Anathswarmy, Anil, 'Reality's Last Stand'. *New Scientist*, Issue 3204, November 17, 2018.

Ancient Aliens, 'The Wisdom Keepers', TV programme, Sept 2017. www.history.co.uk/shows/ancient-aliens

Andersen, Hans Christian, *Classic Fairy Tales*. Dugald Stewart Walker and Hans Tegner (illustrators). Barnes & Noble Inc., new edition, 2015 (first published 1835).

Anderson, Olaf, *Living Water: Viktor Schauberger and the Secrets of Natural Energy*, 1966. Gill books, second edition, 2002.

Bailey, Alice A., *The Consciousness of the Atom*, 1922; *A Treatise on Cosmic Fire*, 1925; *The Soul and Its Mechanism*, 1930. Republished by The Lucis Trust. www.lucistrust.org

Barker, Cicely Mary, *Flower Fairies of the Spring*, 1923. Republished by Warne, 2018.

Bauval, Robert and Gilbert, Adrian, *The Orion Mystery: Unlocking the Secrets of the Pyramids*. Arrow, 1994.

Berman, Bob and Lanza, Robert, *Biocentrism: How Life and Consciousness Are the Keys to Understanding the True Nature of the Universe*. BenBella, 2010.

Bird, Christopher, 'Is the Earth a Large Crystal?', *New Age Journal*, May 1975, pp.36-41. See also http://vortexmaps.com/chris-bird.php

Blackmore, Susan, 'Abduction by Aliens or Sleep Paralysis?' Skeptical Inquirer, Vol. 22, No. 3, May/June 1998. https://skepticalinquirer.org/1998/05/abduction_by_aliens_or_sleep_paralysis/

Blake, William, *Songs of Innocence and of Experience*, 1794. Tate Publishing, illustrated reprint, 2006.

Blavatsky, Helena P., *Isis Unveiled*, 1877. Quest Books, 1997.

Blumrich, Josef F., *The Spaceships of Ezekiel*. Corgi, 1974.

Carson, Rachel, *Silent Spring*, 1962. Penguin Classics, new edition 2000.

Collins, Andrew, *LightQuest: Your Guide to Seeing and Interacting with UFOs, Mystery Lights & Plasma Intelligences*. Eagle Wing Books, 2012.

Cotterell, Maurice, *The Lost Tomb of Viracocha: Unlocking the Secrets of the Peruvian Pyramids*. Headline Book Publishing, 2001.
Cress, Donald A., *René Descartes: Discourse on the Method*, 1637. English translation, Hackett Publishing Co. Inc., third edition, 1998.
Currivan, Jude, *The Cosmic Hologram: In-formation at the Center of Creation*. Inner Traditions, 2017.
Daniel, Alma, Wyllie, Timothy and Ramer, Andrew, *Ask Your Angels*. Ballantine, 1992.
Darwin, Charles, *On the Origin of Species*. John Murray, 1859. http://darwin-online.org.uk/converted/pdf/1861_OriginNY_F382.pdf
Dawkins, Peter, 'How does pilgrimage help the Earth?' The Gatekeeper Trust. www.gatekeeper.org.uk
de Jong, Jan Peter, 'The Cosmogony of the Three Worlds'. www.janpeterdejong.weebly.com
Douglass, Scott, 'Donald Scott: Parker Solar Probe and the Electric Sun', *Space News*, December 14, 2019. https://www.thunderbolts.info/wp/2019/12/14/donald-scott-parker-solar-probe-and-the-electric-sun-space-news/
Dreamland Resort, *Secrets of Area 51 Revealed*, 'FAQ: What other names are used for Area 51?' http://www.dreamlandresort.com/faq/faq_other_names.html
Dyer, Wayne W. Dr., and Garnes, Dee, *Memories of Heaven: Children's Astounding Recollections of the Time Before they Came to Earth*. Hay House, 2015.
Eagleman, David, *The Brain: The Story of You*. Canongate Books, 2015.
Edwards, Frank, *Stranger than Science*. Pan, London, 1963 (first US edition, 1959).
Einstein, Albert, *Relativity: The Special and the General Theory*, 1920. Reprinted by Martino Fine Books, 2010. See also http://www.einstein-online.info/elementary/cosmology.html
Einstein, Albert, *Albert Einstein and the Fabric of Time*, 'The People of Timelessness'. http://everythingforever.com/einstein.htm
Findhorn Foundation. www.findhorn.org
Fortune, Dion, *The Cosmic Doctrine*, 1966. Weiser, 2003.
Gamble, Patrick, psychic artist and visionary. www.patrickgamble.co.uk
Gates, Josh, 'The Rendlesham Forest Incident', *Expedition Unknown*. https://www.facebook.com/watch/?v=10155773406543851

Grimm, Jacob and Wilhelm, *The Original Folk and Fairy Tales of the Brothers Grimm: The Complete First* Edition, 1812. Reprint edition, Andrea Dezso (illustrator), Jack Zipes (translator), Princeton University Press, 2016.

Harris, Chris, Dowser, personal communication, 2018. https://chrisharris.ucoz.com/index/dowsing/0-126

Haywood, Sarah, personal communication, 2018. www.aspire2bfree.com/

Helliwell, Tanis, *Summer with the Leprechauns*, Tanis Helliwell Corporation, 2011.

Holy Bible, *Genesis* 6:4 (American Standard Version); *Ezekiel* 1:5-10, *Luke* 1:30-32 (King James Version).

Jung, Carl Gustav, *The Archetypes and the Collective Unconscious*. Routledge (second edition), 1991.

Kriyananda, Swami, *The Essence of the Bhagavad Gita – Explained by Paramhansa Yogananda*. Crystal Clarity, US (second edition), 2008.

Kyriacou, Christian, *The House Whisperer*. Ki Signature Books, 2014.

László, Ervin, *Science and the Akashic Field: An Integral Theory of Everything*. Inner Traditions, second revised edition, 2007.

Lefors Clark, Richard, Ph.D., 'Diamagnetic Gravity Vortexes', *Anti-Gravity and the World Grid*. David Hatcher Childress (Ed.), Adventures Unlimited Press, 1987.

Lock, Judith, Dowser, 2018. http://healthwellbeing.focusonuk.co.uk/using-dowsing-holistically/

Lovelock, James, *Gaia: A New Look at Life on Earth*. OUP, 1979.

Lovelock, James, *The Revenge of Gaia*. Penguin, 2006.

Matthews, Marian, *Aspects of Reality – A User's Guide to the Universe*. Archive Publishing, 2012.

McCrae, Morrice, 'James Hutton's Theory of the Earth..., 1785', *Journal of the Royal College of Physicians*, Edinburgh, 2012; 42:87–9. https://www.rcpe.ac.uk/sites/default/files/exlibris_2.pdf

McDermott, Alicia, 'Puma Punku: This Ancient Andean Site Keeps Everyone Guessing', *Ancient Origins*, 14 June 2019 - 21:49. www.ancient-origins.net/ancient-places-americas/puma-punku-002

Miller, Hamish, and Broadhurst, Paul, *The Sun and The Serpent*. Penwith Press, 1987.

Montgomery, Ruth, *Strangers Among Us – Enlightened Beings from a World to Come*. New York, Coward, McCann & Geoghegan, 1979.

Moody, Raymond A., Dr, with Perry, Paul, *Life Before Life: Regression into Past Lives*. Pan Books, 1991.
Moody, Raymond A., Dr, *Life After Life*. Rider, 2001 (25th anniversary edition).
Morfitt, Tracy Anne, personal communication, 2018. Public Facebook Group: 'Where Do I Fit Into the Jigsaw we Call Life'.
Musk, Elon, Radio Motherboard podcast, 2 June 2016.
https://motherboard.vice.com/en_us/article/8q854v/elon-musk-simulated-universe-hypothesis
Nelson, Nichols R., *Paradox: a round trip through the Bermuda Triangle*. Dorrance & Company, 1980.
Newton, Isaac, *Philosophiae Naturalis Principia Mathematica* (1687), Stanford Encyclopaedia of Philosophy, 2007.
https://plato.stanford.edu/ entries/newton-principia/
Oxford English Dictionary, Concise, 12th edition. OUP, 2011.
Planck, Max, Science Quotes.
https://todayinsci.com/P/Planck_Max/ PlanckMax-Quotations.htm
Plato, *Allegory of the Cave*.
www.mesacc.edu/~barsp59601/text/lex/defs/a/allegoryofthecave.html
Pogačnik, Marko, *Nature Spirits & Elemental Beings*. Findhorn Press Ltd., 1997.
"Primal", personal communication, 2016.
Randle, Kevin D., *A History of UFO Crashes*. Avon Books, 1995.
Rendlesham Forest Incident, The.
www.therendleshamforestincident.com/
Russell, David, personal communication, 2015. Facebook: 'The Global Brain Project', weekly meditation for world peace.
Russell, Peter, *The Global Brain: The Awakening Earth in a New Century*. Third edition, Floris, 2007.
Sabini, Meredith (Editor), *The Earth has a Soul: C. G. Jung on Nature, Technology & Modern Life*. North Atlantic Books, 2002.
Sambhava, Padma (Compiler), Thurman, Robert (Translator), *The Tibetan Book of the Dead: The Great Book of Natural Liberation Through Understanding in the Between*. Bantam, 1993.
Sanderson, Ivan T. with Hatcher Childress, David, *Invisible Residents: The Reality of Underwater UFOs*. Adventures Unlimited Press, 2005.
Schmitt, Donald R., *Cover-Up at Roswell: Exposing the 70-Year Conspiracy to Suppress the Truth*. New Page Books, 2017.

Scranton, Laird, *The Science of the Dogon: Decoding the African Mystery Tradition*. Inner Traditions, 2006.

Seifer, Marc J., *Wizard: The Life and Times of Nikola Tesla – Biography of a Genius*. Citadel Press (reprint edition), 1996.

Sellers, Josephine, *The Return of Yesterday's People*. Capall Bann, 2002.

Sellers, Josephine, *Parallel Worlds*. Archive Publishing, 2011.

Sitchin, Janet (Editor), *Annunaki Chronicles: A Zecharia Sitchin Reader*. Bear & Company, 2015.

Spencer, John and Anne, *Fifty Years of UFOs: from Distant Sightings to Close Encounters*. Boxtree Ltd., 1997.

Strieber, Whitley, *Communion: A True Story – Encounters with the Unknown*. Arrow Books Ltd., 1988.

Temple, Robert, *The Sirius Mystery: New Scientific Evidence for Alien Contact 5,000 Years Ago*. Inner Traditions Bear & Co., 1998.

Thomas, Lewis, *The Lives of a Cell – Notes of a Biology Watcher*. Viking Press, 1974.

Tolkein, J. R. R., *The Hobbit*, 1937. Second edition, London, George Allen & Unwin, 1951.

Tomas, Andrew, *We are Not the First: Riddles of Ancient Science*. Sphere, New edition, 1972.

Tucker, Jim B., Dr, *Life Before Life: A scientific investigation of children's memories of previous lives*. Piaktus, 2009.

Van Gelder Kunz, Dora, *The Real World of Fairies: A First Person Account*. Quest Books, 1999.

Von Däniken, Erich, *Chariots of the Gods?* Souvenir Press, 1969.

Von Däniken, Erich, *According to the Evidence*. Souvenir Press, 1977.

Washington, George, 'George Washington's Vision and Prophecy About America'. https://sign.org/articles/george-washingtons-vision-and-prophecy-about-america

Watkins, Alfred, *The Old Straight Track*, 1925. Abacus, 1988.

Watts, Barry, *UFOs Down Under: Australasian Encounters*. Amazon, 2017.

Weiss, Brian, Dr., *Same Soul, Many Bodies*. Piatkus, 2004.

Wilcock, David, 'Becker-Hagens – The Global Grid Solution'. https://www.bibliotecapleyades.net/mapas_ocultotierra/esp_mapa_ocultotierra_15.htm

Witteveen, Willem, *The Great Pyramid of Giza: A Modern View on Ancient Knowledge*. Adventures Unlimited Press, 2016.

www.ingramcontent.com/pod-product-compliance
Lightning Source LLC
Chambersburg PA
CBHW052048070526
44584CB00017B/2094